T0396044

Membrane-Distillation in Desalination

Membrane-Distillation in Desalination

Farid Benyahia

CRC Press is an imprint of the
Taylor & Francis Group, an **informa** business

CRC Press
Taylor & Francis Group
6000 Broken Sound Parkway NW, Suite 300
Boca Raton, FL 33487-2742

Printed on acid-free paper

International Standard Book Number-13: 978-1-4987-4854-4 (Hardback)

Library of Congress Cataloging-in-Publication Data

Names: Benyahia, Farid, author.
Title: Membrane-distillation in desalination / Farid Benyahia.
Description: New York, NY: CRC Press/Taylor & Francis Group, 2019. |
Includes bibliographical references and index.
Identifiers: LCCN 2019015119 | ISBN 9781498748544 (hardback: acid-free paper)
| ISBN 9781315117553 (ebook)
Subjects: LCSH: Saline water conversion.
Classification: LCC TD480.4 .B46 2019 | DDC 628.1/674--dc23
LC record available at https://lccn.loc.gov/2019015119

Visit the Taylor & Francis Web site at
http://www.taylorandfrancis.com

and the CRC Press Web site at
http://www.crcpress.com

This book is dedicated to the millions of people on this planet denied their human right to water and sanitation (UN General assembly resolution 64/292, July 2010).

Contents

Preface

Membrane distillation for desalination is a very popular topic for research with graduate students and research staff in general, worldwide. It is an emerging technology for water treatment, especially for desalting seawater and high salinity industrial waste streams using low carbon energy sources such as low-grade waste heat or solar power. It is a technology that can be deployed to remote places where communities suffer from water shortages and access to clean freshwater from the main water supply is not possible or prohibitively expensive. However, ever since its development in the nineties, its potential to produce ultra-high quality for industry where it can easily be thermally integrated to exploit low-grade waste heat has been recognized. Yet, some 30 years on we have yet to see this marvelous technology deployed on a commercial scale. The reasons are diverse: some are purely technical and related to the stability of the hydrophobic membranes used and others are related to the economics of a technology that was not given the chance it rightly deserves to be launched on a large scale. However, the massive growth in interest in membrane technology is encouraging, and many are hopeful that it will be deployed in the not too distant future owing to the impressive amount of work done on the topic. This book is a concise reference book rich in content and literature references suitable for a wide range of readers: new graduate students, new research assistants, university professors and laboratory technicians. For such a diverse audience, the style of the write-up and depth/breadth of content was very carefully selected. It does not go too deep in every topic but provides the most salient issues in good detail, providing ample references for those who wish to dig deeper into the various topics. The book also contains an important introduction chapter on the current water situation worldwide where water-stressed regions are no longer the low GDP nations but almost every nation on Earth regardless of their degree of affluence. This book, in its first edition, will be the friendly companion of young researchers who wish to invest their future career on low carbon water technologies.

Author

Farid Benyahia is a professor of chemical engineering and fellow of the institution of chemical engineers, currently affiliated with the school of chemical engineering at the University of Birmingham in the United Kingdom. He held previously positions of faculty member and head of department in the Gulf and was prior to that senior lecturer in the United Kingdom. He was educated at the Universities of Newcastle and Aston in the United Kingdom. His research interests include low carbon desalination, carbon management, concentrated brine management, contaminated soils and water treatment, and multiphase reactor systems.

1

The Water Nexus and Desalination

1.1 Introduction: The Twenty-First Century Context for the Pursuit of Sustainable Water Resources

Humanity embarked on the twenty-first century facing a number of challenges leaving politicians, visionaries and academics pondering on elusive solutions for a long time to come. Some such major challenges include water resources, energy, food security and population. This is by no means an exhaustive list. However, the four challenges mentioned are intimately related. By far, availability and access to clean freshwater constitute the major concern to decision makers globally. Indeed, the inexorable increase in the world population over the past decades put a considerable strain on Earth's freshwater. While planet Earth has an immense salt water inventory in the form of oceans and seas, freshwater constitutes a mere 2.5% of the global total water and from this fraction, nearly 69% of freshwater is locked as frozen glaciers and permanent icecaps, leaving just over 30% of freshwater available as groundwater and 0.3% of freshwater available as surface water. Surface freshwater constitutes a tiny 0.3% of freshwater available and the remaining 0.9% are accounted for as soil moisture, permafrost and swamps. This remarkable and striking estimate reported by Igor Shiklomanov [1] still remains a valuable reference for the global freshwater inventory of planet Earth. Figure 1.1 depicts Earth's global water and freshwater distribution.

Unlike other natural resources, water circulates naturally according to the global hydrological cycle offering the possibility to recharge natural and man-made water catchments. However, according to Oki and Kanae [2], it is the flow of water that should be the focus rather than the stock in the assessment of water resources. Oki and Kanae [2] also pointed out that due to the long time it may take to recharge groundwater reservoirs naturally to their original volume stored, if ever, groundwater has sometimes been called "fossil water."

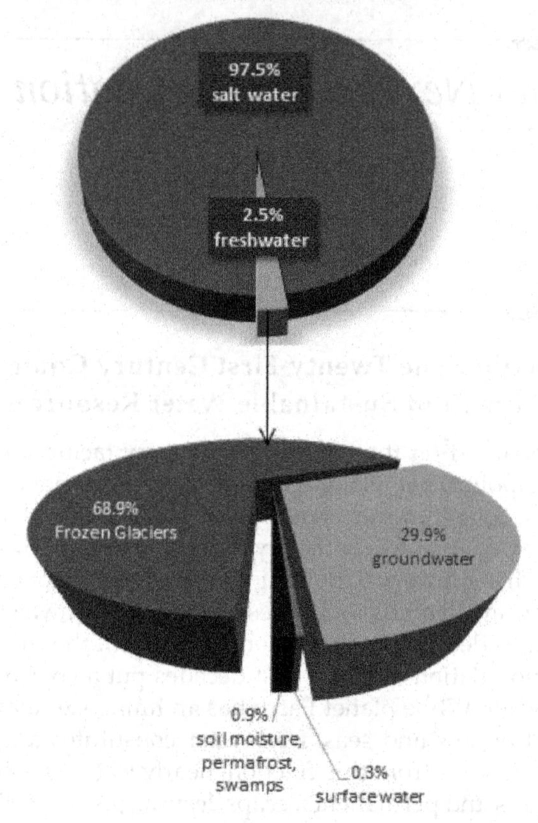

FIGURE 1.1
Earth's global water and freshwater distribution.

This is particularly significant when groundwater is being withdrawn in many parts of the world at a rate that exceeds natural recharging [3–5] and that globally, about 70%–80% of the total water consumed is used in agriculture [1,6].

The surge in freshwater consumption is primarily due to an increasing world population that needs food from agricultural activity, industrial activities and urban/rural development to maintain or improve its lifestyle. The most recent United Nations population census [7] depicts an unprecedented upward trend in the past decades, reaching 7.63 billion in 2017 and projected to reach over 10 billion past the 2050 horizon. Figure 1.2 shows the world

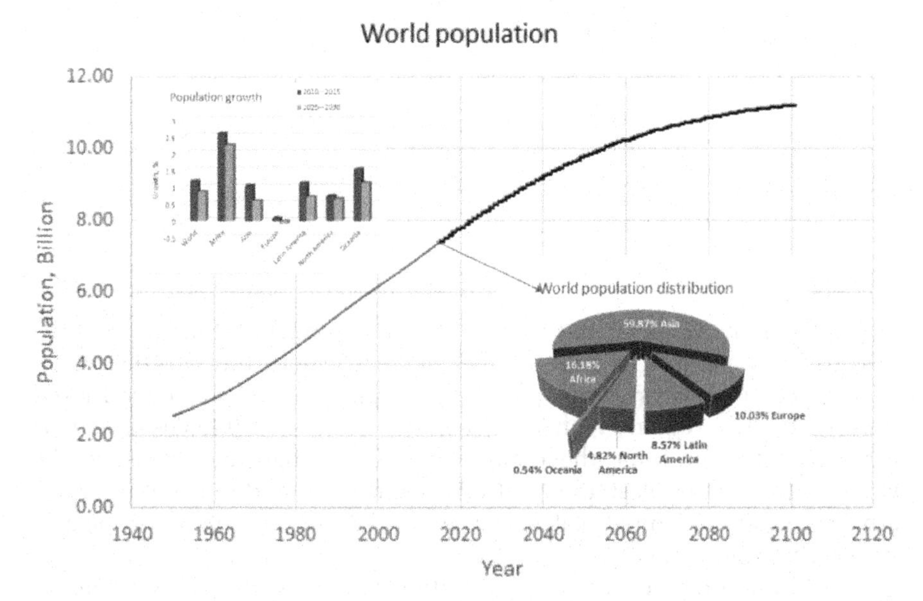

FIGURE 1.2
The world population trend from the 1950s to the 2100 horizon.

population trend from the 1950s to the 2100 horizon. It can be seen that the major fraction of the world population is currently (2017) located in Asia (59.87%), followed by Africa (16.18%), Europe (10.03%), Latin America (8.57%), North America (4.82%) and finally Oceania (0.54%). The population growth in Figure 1.2 indicates that the global population will continue an upward trend with the exception of Europe where a negative growth may be experienced past the 2025 horizon. Because the uneven distribution of renewable freshwater reserves (RFWR) has been somewhat exacerbated by the climate system, most regions in the world, regardless of the status of economic development or wealth, have been affected by water shortages intermittently or permanently. Consequently, a measure of water scarcity has been put forward in the form indices commonly called water scarcity indices. These indices have been reviewed by Brown and Matlock [8]. However, the simplest water scarcity index covered by Oki and Kanae [2] has been used to map water-stressed regions in the world. This water scarcity index is represented by the following equation:

$$R_{ws} = \frac{W - S}{Q} \tag{1.1}$$

TABLE 1.1
Water Scarcity Scale

$R_{ws} < 0.1$	No water stress
$0.1 \leq R_{ws} < 0.2$	Low water stress
$0.2 \leq R_{ws} < 0.4$	Moderate water stress
$R_{ws} \geq 0.4$	High water stress

where R_{ws} is the water scarcity index, W is the annual abstraction, S is the desalinated water and Q is the annual available water. From this definition, a water scarcity scale system [9] has been proposed, as seen in Table 1.1 and used to identify regions with R_{ws} equal or greater than 0.4. These regions include northern China, the region bordering Indian and Pakistan, the whole of the Middle-East and North Africa (MENA) and much of the western regions of the United States of America. Parts of southern Europe also suffer from water scarcity.

Tables 1.2 and 1.3 show the water withdrawal by source and usage sector in the MENA region and water withdrawal by sector (including desalination) as % of total, respectively [6].

Table 1.2 shows that the Arabian Gulf region has one of the highest groundwater consumption in the world for agriculture while Africa has the highest proportion of its freshwater resource used for agriculture. The world average for water usage in agriculture is 69%. The large water consumption for agriculture and indirectly for food production in all forms led to the introduction of a particular footprint called "water footprint" (akin to the carbon footprint) [10–12]. Ironically, most healthy foods (vegetables) have a rather large water footprint [13,14]. This has profound implications on the meaning of sustainability of freshwater supplies. In other words, water security is a global issue and truly deserves the phrase "water nexus" [15–18].

Given the status of the water scarcity globally, how can we increase the water supply when the hydrological water cycle is not only finite, but also exacerbated by an increasing world population and an erratic climate pattern that does not spare any region in the world?

TABLE 1.2

Water Withdrawal by Source and Usage Sector in MENA Region

	Source (% of Total)				Usage Sector (% of Total)			
Country	Groundwater	Desalinated Water	Treated Wastewater	Surface Water	Irrigation + Livestock	Municipalities	Industry	Total Withdrawal (km^3)
KSA	90.3	4.4	0.7	4.6	88	9	3	23.666
UAE	70	24	6	0	83	15	2	3.998
Kuwait	45	46	9	0	54	44	2	0.913[1]
Qatar	49	41	10	0	59	39	2	0.444[2]
Oman	89	8	3	0	89	10	1	1.321[3]
Bahrain	66	29	5	0	45	49	6	0.357[3]
Egypt	10.3	0.3	5.4	84	86	11	3	78.00[5]
Algeria	35.6	7.3	0.1	57	59	36	5	8.425[4]
Tunisia	62	1	2	35	80	15	5	3.305[6]
Morocco	21.80	0.07	0.66	77.47	88	10	2	10.65[5]
Jordan	59	1	9	31	65	31	4	0.941[5]
Israel	80[a]	7	13	—	58	36	6	1.954[7]

[1](2002) [2](2005) [3](2003) [4](2012) [5](2010) [6](2011) [7](2004)

[a] Combined ground and surface.

TABLE 1.3
Water Withdrawal by Sector (including Desalination) as % of Total

	Agriculture	Municipal	Industry
World	69	19	12
Africa	81	15	4
Americas	48	14	37
Asia	81	9	10
Europe	25	21	54
Oceania	65	20	15
Middle-East	84	9	7
North Africa	84	13	3
Sub-Sahara	79	16	5

1.2 Increasing Water Supply

This may be achieved from the hydrological cycle (rain capturing infrastructures), water reuse (treatment and recycling of wastewater; some parts of the world call wastewater used water for social reasons) and seawater desalination. However, there are constraints for any potential solution, be it economic, environmental or social. It is legitimate to ask about the sustainability of augmented freshwater supply: indeed, energy and cost are major determining factors.

Where possible and economically more advantageous, wastewater treatment and reuse would be more favorable [19–21]. However, there are a number of constraints that must be taken into account (such as social sensitivities, health risks and regulations) [22–27] and hence desalination of saline water must be considered as a viable alternative in countries with sea coasts.

The relative advantages of desalination: it is virtually climate change independent, and salty water (oceans, seas) is abundant and can be considered as a sustainable resource when wisely exploited. However, there are disadvantages of desalination: effects on the ecosystem (potential local salinity increases if rejects are not dispersed sufficiently, damage to marine life at intake if no precautions taken), energy intensity and carbon emissions (if fossil fuels are used). Some progress has been made to minimize effects on the ecosystem, and this will be briefly highlighted in later sections.

1.3 Exploiting the Vast Salty Water Resources: Desalination

The first thermal desalination facility was established in Great Britain after a patent was granted in 1869 [28]. The freshwater produced was mainly used to supply ships. In the same period, the first major issues related to scaling were reported [28]. Since then, desalination technologies evolved and spread to all parts of the globe where freshwater is scarce and seawater (or brackish) water is available, becoming a multibillion-dollar industry. In 2017, the global online cumulative desalination capacity reached 92.5 million m^3/day (99.8 million m^3/day contracted) with about 19,372 plants around the world (compared to 18,983 in 2016) [29]. Both capacity and number of plants continue to grow as global demand in water-stressed regions surges with increased population and associated needs.

Figure 1.3 depicts the trend of annual online and contracted global desalination capacity (1980–2016). It also shows that in the period 1990–2016, membrane-based desalination dominated global capacities. Over that period (1990–2016), membrane-based desalination accounted for 79.88% of the capacity compared to 20.12% for thermal desalination. Much of the global thermal desalination capacity is to be found in the Arabian Gulf where legacy

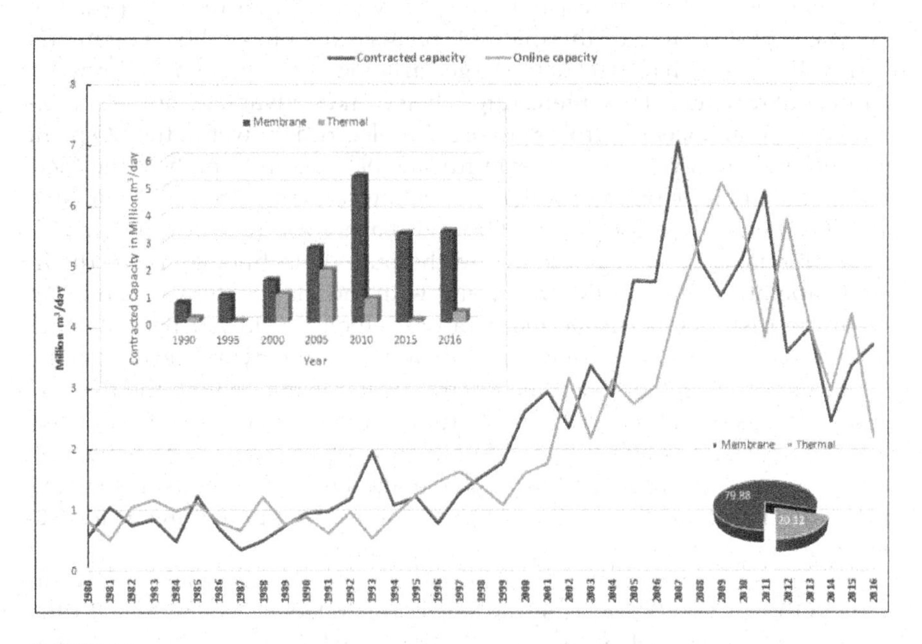

FIGURE 1.3
The trend of annual online and contracted global desalination capacity (1980–2016).

thermal plants built 20–40 years ago continue to function with good maintenance. These plants built with expensive cupronickel high alloys indeed have a long life and can operate for quite some time in the future with occasional re-tubing. In the Arabian Gulf region, desalination plants tend to be part of the so-called power and water "co-generation" schemes to improve overall energy efficiency.

Modern thermal desalination technologies include multistage flash distillation (MSF) and multi-effect distillation (MED) with thermal vapor compression (TVC) coupled with power generation where heat is recovered from the steam turbine blowdown or gas turbine exhaust (to generate steam) and used to preheat the feed for the MSF/MED distillation plants. Despite significant improvements in the design of thermal desalination plants, the specific energy consumption is still high [30] compared to reverse osmosis. These topics have been covered considerably in terms of thermodynamics, design and economics [31–39] including fouling/scaling mitigation [40–44]. The environmental impacts of desalination plants have been reported in a number of articles [45–51] and these involved a range of concerns such as carbon emissions and marine ecosystems.

The most recent thermal desalination plants built in the Arabian Gulf enter a new era of "hybrid thermal-reverse osmosis" plants. This topic is not new and has been covered in research previously [52,53] and recently [54]. The advantage of such hybrid "thermal-RO" plants is their flexibility of operation during the year with respect to the large variation of power demand between summer and winter. This interesting scheme takes advantage of the combination of gas and steam turbines to provide electrical power to the RO plant and thermal energy from the steam turbine blowdown to preheat the thermal desalination plant. Freshwater production can range from 100% RO to a split 70% thermal and 30% RO, to 100% thermal according to power demand. An additional advantage of the hybrid thermal-RO technology is the distillate temperature: mixing RO freshwater with thermal freshwater can bring down the distillate-permeate temperature which would not need cooling prior to storage. Another interesting advantage of the hybrid technology is the lower boron content when the RO freshwater is mixed with the thermal freshwater: thermal freshwater has virtually zero boron content. Production capacity of existing older thermal desalination plants can be augmented by revamping and installing an RO plant to make a hybrid plant using existing intake/discharge facilities with minimal engineering-procurement for RO pre-treatment and associated pipe/pump systems.

Examples of recently started hybrid plants in the Arabian Gulf include an MED-RO plant (591,000 m^3/day) in Fujairah, United Arab Emirates. In the kingdom of Saudi Arabia, the world's largest hybrid plant (RO producing 309,360 m^3/day and multi-stage flash evaporation producing 727,130 m^3/day) is located in Ras Al-Khair (production started in 2017) [55]. A planned hybrid MSF-RO plant to be located in Umm al Houl, south of Doha in the state of Qatar with a total production 590,000 m^3 was inaugurated in March 2019.

With over half a century of documented experience in desalination in the scientific and industrial literature, one can legitimately ask what were the main lessons learned and is there room for new desalination technologies? Such a question was asked a couple of decades ago regarding improved thermal desalination, but the world of desalination has changed dramatically [56]. The water nexus and associated topics hint at a complex field of salt water desalination where the nature of energy, energy requirements, design, cost, ecological considerations and social pressures play important roles in defining the future shape of desalination technologies. At the same time, fundamentals of science will always put a framework where energy requirements and efficiencies cannot go below certain limits, though they can be approached through judicious design of systems and deployment of new materials.

1.3.1 Lessons from Desalination Literature and Industrial Practice

One of the most important limits to energy required for saltwater desalination, regardless of the technology adopted, comes from fundamentals of thermodynamics [57,58] in the form of equation (1.2):

$$-d\left(\Delta G_{mix}\right) = -RTlna_w dn_w = \pi \bar{V}_w \, dn_w \tag{1.2}$$

where ΔG_{mix} is the free energy of mixing, R is the ideal gas constant, T is the absolute temperature, a_w is the activity of water, n_w is the number of moles of water, π is the osmotic pressure of the seawater and V_w is the molar volume of water. By integrating equation (1.2), it turns out that for a typical saltwater of 35,000 ppm and an ideal reversible thermodynamic system, assuming 0% recovery (just overcoming the solution osmotic pressure), a minimum of 0.76 kWh/m^3 is required, while for the classic 50% recovery, the minimum required would be 1.06 kWh/m^3. These reference values are excellent limits no one can challenge regardless of the method to carry out the separation. Elimelech and Phillip [58] argued that another useful practical limit on the minimum energy required for RO assuming no concentration polarization, 100% efficiency for pumps and energy recovery exchangers, then another "practical limit" of 1.56 kWh/m^3 is prescribed. The difference between the ideal and practical limits estimated as 0.50 kWh/m^3 is attributed to the fact that the system has a finite size and does not operate in a reversible thermodynamic fashion.

How do current reverse osmosis plants fare compared to the above theoretical limits?

There is abundant literature on energy consumption in desalination plants [59–63]. For desalination of seawater of typical concentration 35,000 ppm, a range of 2.5–4.0 kWh/m^3 is often quoted in the literature for RO. There is a range because not all reverse osmosis plants have the same specification, like number of RO stages, feed salinity, extent of pretreatment and of course the technology used for reject brine energy recovery (pressure exchange, not thermal). A typical modern RO plant with good energy recovery technology can have energy requirements shown in Table 1.4. Data in Table 1.4 were extracted from Voutshkov's article [62]. Clearly, the most significant energy consumed is to be found in the reverse osmosis operation. In an interesting recent article, Karabelas et al. [59] further broke down the RO step energy consumption into several constituent losses and resistances. These are shown in Table 1.5. The membrane resistance is the second important energy consumption after overcoming the saline solution osmotic pressure, followed by the energy consumed in the pressure recovery device. It is precisely in the membrane resistance and the pressure recovery that reverse osmosis made significant energy consumption reductions since the 1970s when the RO specific energy consumption was significantly higher, typically over 10 kWh/m³ and gradually reducing to the current values (2–4 kWh/m³) through membrane improvements and more efficient pressure exchangers (energy recovery devices) [63].

In Karabelas et al.'s data, the total RO step specific energy is around 2.4 kWh/m³. This value is close to the value of 2 kWh/m³ indicated by Elimelech and Philip [58] and Lee et al. [63] for the best performance yet.

From the extensive literature on reverse osmosis, it becomes clear that much of the improvements on performance (salt rejection) and reduced

TABLE 1.4

Energy Requirement Breakdown in a Typical Modern RO Plant

Operation	Energy Consumed (kWh/m³)	Fraction of Energy Consumed (%)
RO process	2.54	71
Pre-treatment	0.39	10.8
Product delivery	0.18	5.0
Intake	0.19	5.3
Other facilities	0.27	7.6

TABLE 1.5

RO Step Energy Requirement Breakdown

	Energy Consumed (kWh/m³)	Fraction of Energy Consumed (%)
Osmotic pressure	1.2	50.55
Membrane resistance	0.574	24.18
Retentate friction loss	0.057	2.40
Permeate friction loss	0.0012	0.05
Concentration polarization	0.057	2.40
Energy recovery device	0.485	20.42

specific energy consumption in reverse osmosis is due to the development of the thin film composite (TFC) polyamide membranes consisting of three layers: a polyester layer acting as structural support (150 μm thick), a microporous interlayer (40 μm) and an ultra-thin barrier layer on the upper surface (0.2 μm) with pores less than 0.6 nm [63]. The TCF polyamide membrane is still the "gold standard" in RO. The most widely used membrane geometry is the spiral wound, accounting for over 90% of the market share [63]. Improvements in energy recovery from the high pressure concentrated retentate also contributed to the reduction in specific energy consumption. As far as RO process design is concerned, multistage operations and multistage closed circuit have been described as means to approaching the thermodynamic limit of energy consumption [64].

1.4 Improved Environmental Impacts on the Ecosystem

Seawater intake for desalination, in particular the surface intake, has been the subject of scrutiny for its impact on marine life. Modern plants now use the so-called sub-surface intakes to minimize the impact on marine life [65–67].

New membranes: Focus should be on fouling resistance instead of energy

Beyond the improvements mentioned previously, what could be done to improve RO desalination further? The literature has abundant coverage of membrane modification [68,69] as well as new exotic materials such as graphene [70–72] and aquaporin [73,74] for desalination. However, one should be cautious about disproportionate claims in specific energy consumption reduction since we are operating close to the thermodynamic limit. New material or modified existing material used in RO membranes can, however, be used to reduce fouling [68,71,75–80] and perhaps slightly reduce the membrane resistance provided the salt rejection is maintained. New membranes that reduce fouling (and all negative aspects like increasing pumping pressure to maintain productivity, chemical cleaning, cost of cleaning chemicals, effect of chemicals on membrane life) will make a positive impact on reverse osmosis economics.

1.5 Prospects of Solar Energy in Desalination

We have seen that desalination is an energy intensive process with a minimum energy requirement prescribed by thermodynamic fundamentals of saline solutions. So far, fossil fuels are the primary fuels used to power desalination plants and the despite recent improvements, the carbon footprint of such industry remains even after phasing out thermal processes. In this respect, solar energy will have an increasing role in future desalination plants, especially in parts of the world that suffer from water stress. Fortunately, these regions also enjoy a great deal of sunshine and have communities that are remote from the comfort of large cities and industrial zones [81,82]. There are good prospects for solar energy for desalination and some studies have been reported [83–85]. It appears that one of the most important factors in using solar and indeed any renewable source of energy is cost and affordability. This is consistent with Resolution 64/292 of the United Nations General Assembly (2010) that recognizes the human right to water and sanitation [86]. Many parts of the world suffering from water stress are also poor. This direction seems appropriate and more work is needed in that direction.

1.6 Concluding Remarks

This chapter was centered around the global water nexus. Global demand for freshwater is increasing while many parts of the world are suffering from water stress regardless of their status of economic development or wealth. The climate change exacerbated natural hydrological cycle may not provide sufficient, regular and predictable water supplies to increasing parts of the world. There is an urgent need to augment water supply in these regions. While water reuse may be the logical and preferred option, it may not be sufficient nor would it be acceptable for some water uses for social reasons in some parts of the world. Desalination of seawater is therefore seen as an unavoidable option in many parts of the world to provide a supply of freshwater to sustain economic development and perhaps reduce depletion of valuable groundwater reserves. Desalination of seawater for freshwater supply is virtually climate change independent.

Seawater desalination is a mature industrial practice and much of the underlying science is known. However, there is much to do in terms of technological advancements to reduce energy consumption, overcome operational problems in their diverse forms, reduce environmental impact and cost of saltwater desalination.

A gradual shift to renewable energy to power desalination plants, and in particular solar energy, is already appearing in some parts of the world [87] (UAE, USA, and Morocco), and the trend is definitely upward in years to come.

A membrane-based desalination method that is emerging but not commercially deployed yet is membrane distillation. It is a fast-growing research topic and the main focus in this book. One of its distinctive advantages is the utilization of low-grade heat and solar energy to power the process. This feature is particularly attractive when this emerging technology is deployed in remote, off-grid communities or thermally integrated with existing chemical plants that discharge a large amount of low-grade heat to the environment as part of the cooling processes. The next chapters will be dedicated to exploring this promising desalination technology, building on the extensive knowledge and experience traditional desalination technologies provide us.

References

1. Shiklomanov, I.A., *World Water Resources*. 1998, Paris, France: United Nations Educational, Scientific and Cultural Organization: International Hydrological Programme.
2. Oki, T. and S. Kanae, Global hydrological cycles and world water resources. *Science*, 2006. **313**(5790): pp. 1068–1072.
3. Postel, S.L., G.C. Daily, and P.R. Ehrlich, Human appropriation of renewable fresh water. *Science*, 1996. **271**(5250): pp. 785–788.
4. Postel, S.L., Entering an Era of water scarcity: The challenges ahead. *Ecological Applications*, 2000. **10**(4): pp. 941–948.
5. Hoekstra, A.Y. et al., Global monthly water scarcity: Blue water footprints versus blue water availability. *PLOS One*, 2012. **7**(2): pp. e32688.
6. Aquastat, F., Food and Agriculture Organization of the United Nations (FAO). http://www.fao.org/statistics/en. Accessed on February 15, 2018. 2016, United Nations.
7. United Nations, Department of Economic and Social Affairs, Population Division, *World Population Prospects: The 2017 Revision*, DVD edn. 2017.
8. Brown, A. and M.D. Matlock, *A Review of Water Scarcity Indices and Methodologies*. Fayetteville, AK: University of Arkansas. 2011.
9. Oki, T. et al., Global assessment of current water resources using total runoff integrating pathways. *Hydrological Sciences Journal*, 2001. **46**(6): pp. 983–995.
10. Lovarelli, D., J. Bacenetti, and M. Fiala, Water footprint of crop productions: A review. *Science of the Total Environment*, 2016. *548–549*: pp. 236–251.
11. Hoekstra, A.Y., A critique on the water-scarcity weighted water footprint in LCA. *Ecological Indicators*, 2016. **66**: pp. 564–573.
12. Hoekstra, A.Y. and A.K. Chapagain, Water footprints of nations: Water use by people as a function of their consumption pattern. *Water Resources Management*, 2007. **21**(1): pp. 35–48.
13. Mekonnen, M.M. and A.Y. Hoekstra, The green, blue and grey water footprint of crops and derived crop products. *Hydrology and Earth System Sciences*, 2011. **15**(5): pp. 1577–1600.
14. Mekonnen, M.M. and A.Y. Hoekstra, Water footprint benchmarks for crop production: A first global assessment. *Ecological Indicators*, 2014. **46**: pp. 214–223.
15. Dai, J. et al., Water-energy nexus: A review of methods and tools for macro-assessment. *Applied Energy*, 2018. **210**: pp. 393–408.
16. Cai, X. et al., Understanding and managing the food-energy-water nexus—Opportunities for water resources research. *Advances in Water Resources*, 2018. **111**: pp. 259–273.

17. Laio, F., M.C. Rulli, and S. Suweis, The challenge of understanding the water-food nexus complexity. *Advances in Water Resources*, 2017. **110**: pp. 406–407.
18. Rao, P. et al., Technology and engineering of the water-energy nexus. *Annual Review of Environment and Resources*, 2017. **42**(1): pp. 407–437.
19. Wilcox, J. et al., Urban water reuse: A triple bottom line assessment framework and review. *Sustainable Cities and Society*, 2016. **27**: pp. 448–456.
20. Yi, L. et al., An overview of reclaimed water reuse in China. *Journal of Environmental Sciences*, 2011. **23**(10): pp. 1585–1593.
21. Prisciandaro, M. et al., Process analysis applied to water reuse for a "closed water cycle" approach. *Chemical Engineering Journal*, 2016. **304**: pp. 602–608.
22. Elsokkary, I.H. and A.F. Abukila, Risk assessment of irrigated lacustrine & calcareous soils by treated wastewater. *Water Science*, 2014. **28**(1): pp. 1–17.
23. Al Omron, A.M. et al., Long term effect of irrigation with the treated sewage effluent on some soil properties of Al-Hassa Governorate, Saudi Arabia. *Journal of the Saudi Society of Agricultural Sciences*, 2012. **11**(1): pp. 15–18.
24. Bouma, J., Irrigation with treated sewage effluent: A. Feigin, I. Ravina and J. Shalhevet. 1991 Springer-Verlag. 224 pp, DM 228. ISBN-3-540-50804-X. *Agricultural Water Management*, 1992. **20**(4): pp. 343–344.
25. Feigin, A., I. Ravina, and J. Shalhevet, *Irrigation with Treated Sewage Effluent: Management for Environmental Protection*. 1991, New York: Springer-Verlag.
26. Jagals, P. et al., *A Review of the Applicability of the South African Guide for the Permissible Utilisation and Disposal of Treated Sewage Effluent in Agriculture and Aquaculture*. 2002, Gezina, South Africa: Water Research Commission.
27. Jamwal, P. and A.K. Mittal, Reuse of treated sewage in Delhi city: Microbial evaluation of STPs and reuse options. *Resources, Conservation and Recycling*, 2010. **54**(4): pp. 211–221.
28. "Desalination." World of Earth Science. Retrieved February 18, 2018 from Encyclopedia.com., in http://www.encyclopedia.com/science/encyclopedias-almanacs-transcripts-and-maps/desalination.
29. IDA-GWI, *Desalination Yearbook 2017–2018.*, ed. U. Ed., Oxford, UK: Media Analytics, 2017.
30. Al-Karaghouli, A. and L.L. Kazmerski, Energy consumption and water production cost of conventional and renewable-energy-powered desalination processes. *Renewable and Sustainable Energy Reviews*, 2013. **24**: pp. 343–356.
31. El-Dessouky, H. et al., Multistage flash desalination combined with thermal vapor compression. *Chemical Engineering and Processing: Process Intensification*, 2000. **39**(4): pp. 343–356.
32. El-Dessouky, H.T. and H.M. Ettouney, Multiple-effect evaporation desalination systems thermal analysis. *Desalination*, 1999. **125**(1): pp. 259–276.
33. Darwish, M.A. and H. El-Dessouky, The heat recovery thermal vapour-compression desalting system: A comparison with other thermal desalination processes. *Applied Thermal Engineering*, 1996. **16**(6): pp. 523–537.
34. Dahdah, T.H. and A. Mitsos, Structural optimization of seawater desalination: II novel MED–MSF–TVC configurations. *Desalination*, 2014. **344**: pp. 219–227.
35. Lienhard, J.H., M.H. Karan, M.H. Sharqawy, G.P. Thiel, Thermodynamics, exergy, and energy efficiency in desalination systems. *Desalination Sustainability: A Technical, Socioeconomic, and Environmental Approach*, ed. H.A. Arafat. 2017: Elsevier Publishing.

36. El-Dessouky, H.T. and H.M. Ettouney, *Fundamentals of Salt Water Desalination*. 2002, Amsterdam, the Netherlands: Elsevier Science B.V., p. 670.
37. Ghaffour, N., T.M. Missimer, and G.L. Amy, Technical review and evaluation of the economics of water desalination: Current and future challenges for better water supply sustainability. *Desalination*, 2013. **309**: pp. 197–207.
38. Hassan, A.S. and M.A. Darwish, Performance of thermal vapor compression. *Desalination*, 2014. **335**(1): pp. 41–46.
39. Al-Sahali, M. and H. Ettouney, Developments in thermal desalination processes: Design, energy, and costing aspects. *Desalination*, 2007. **214**(1): pp. 227–240.
40. Martynova, O.I. et al., Evaluation of thermal desalination plants water chemistry. *Desalination*, 1983. **47**(1): pp. 63–69.
41. Hodgkiess, T. et al., Acid cleaning of thermal desalination plant: do we need to use corrosion inhibitors? *Desalination*, 2005. **183**(1): pp. 209–216.
42. Budhiraja, P. and A.A. Fares, Studies of scale formation and optimization of antiscalant dosing in multi-effect thermal desalination units. *Desalination*, 2008. **220**(1): pp. 313–325.
43. Al-Rawajfeh, A.E. et al., Scale formation model for high top brine temperature multi-stage flash (MSF) desalination plants. *Desalination*, 2014. **350**: pp. 53–60.
44. Alsadaie, S.M. and I.M. Mujtaba, Dynamic modelling of heat exchanger fouling in multistage flash (MSF) desalination. *Desalination*, 2017. **409**: pp. 47–65.
45. Liu, T.-K., T.-H. Weng, and H.-Y. Sheu, Exploring the environmental impact assessment commissioners' perspectives on the development of the seawater desalination project. *Desalination*, 2018. **428**: pp. 108–115.
46. Darwish, M., A.H. Hassabou, and B. Shomar, Using seawater reverse osmosis (SWRO) desalting system for less environmental impacts in Qatar. *Desalination*, 2013. **309**: pp. 113–124.
47. Tarnacki, K.M. et al., Comparison of environmental impact and energy efficiency of desalination processes by LCA. *Water Science and Technology-Water Supply*, 2011. **11**(2): pp. 246–251.
48. Mezher, T. et al., Techno-economic assessment and environmental impacts of desalination technologies. *Desalination*, 2011. **266**(1–3): pp. 263–273.
49. Lattemann, S. and T. Höpner, Environmental impact and impact assessment of seawater desalination. *Desalination*, 2008. **220**(1–3): pp. 1–15.
50. Elabbar, M.M., The Libyan experimental on the environmental impact assessment for desalination plants. *Desalination*, 2008. **220**(1–3): pp. 24–36.
51. Abu Qdais, H., Environmental impacts of the mega desalination project: The Red–Dead Sea conveyor. *Desalination*, 2008. **220**(1): pp. 16–23.
52. El-Sayed, E. et al., Performance evaluation of two RO membrane configurations in a MSF/RO hybrid system. *Desalination*, 2000. **128**(3): pp. 231–245.
53. Almulla, A., A. Hamad, and M. Gadalla, Integrating hybrid systems with existing thermal desalinationplants. *Desalination*, 2005. **174**(2): pp. 171–192.
54. Bandi, C.S., R. Uppaluri, and A. Kumar, Global optimality of hybrid MSF-RO seawater desalination processes. *Desalination*, 2016. **400**: pp. 47–59.
55. Sanza, M.A., V. Bonnélyea, and G. Cremerb, Fujairah reverse osmosis plant: 2 years of operation. *Desalination*, 2007. **203**(1): pp. 91–99.
56. Al-Gobaisi, D.M.K., A quarter-century of seawater desalination by large multistage flash plants in Abu Dhabi (Plant performance analysis, assessment, present efforts toward enhancement and future hopes). *Desalination*, 1994. **99**(2): pp. 509–512.

57. Spiegler, K.S. and Y.M. El-Sayed, The energetics of desalination processes. *Desalination*, 2001. **134**(1): pp. 109–128.
58. Elimelech, M. and W.A. Phillip, The future of seawater desalination: Energy, technology, and the environment. *Science*, 2011. **333**(6043): pp. 712–717.
59. Karabelas, A.J. et al., Analysis of specific energy consumption in reverse osmosis desalination processes. *Desalination*, 2018. **431**: pp. 15–21.
60. Liu, C., K. Rainwater, and L. Song, Energy analysis and efficiency assessment of reverse osmosis desalination process. *Desalination*, 2011. **276**(1–3): pp. 352–358.
61. Li, M., Reducing specific energy consumption in Reverse Osmosis (RO) water desalination: An analysis from first principles. *Desalination*, 2011. **276**(1–3): pp. 128–135.
62. Voutchkov, N., Energy use for membrane seawater desalination—Current status and trends. *Desalination*, 2018. **431**: pp. 2–14.
63. Lee, K.P., T.C. Arnot, and D. Mattia, A review of reverse osmosis membrane materials for desalination—Development to date and future potential. *Journal of Membrane Science*, 2011. **370**(1–2): pp. 1–22.
64. Lin, S. and M. Elimelech, Staged reverse osmosis operation: Configurations, energy efficiency, and application potential. *Desalination*, 2015. **366**: pp. 9–14.
65. Missimer, T.M. et al., Subsurface intakes for seawater reverse osmosis facilities: Capacity limitation, water quality improvement, and economics. *Desalination*, 2013. **322**: pp. 37–51.
66. Dehwah, A.H.A. and T.M. Missimer, Subsurface intake systems: Green choice for improving feed water quality at SWRO desalination plants, Jeddah, Saudi Arabia. *Water Research*, 2016. **88**: pp. 216–224.
67. Shahabi, M.P., A. McHugh, and G. Ho, Environmental and economic assessment of beach well intake versus open intake for seawater reverse osmosis desalination. *Desalination*, 2015. **357**: pp. 259–266.
68. Zhang, Y. et al., Surface modification of polyamide reverse osmosis membrane with organic-inorganic hybrid material for antifouling. *Applied Surface Science*, 2018. **433**: pp. 139–148.
69. Xu, G.-R., J.-N. Wang, and C.-J. Li, Strategies for improving the performance of the polyamide thin film composite (PA-TFC) reverse osmosis (RO) membranes: Surface modifications and nanoparticles incorporations. *Desalination*, 2013. **328**: pp. 83–100.
70. Zahirifar, J. et al., Fabrication of a novel octadecylamine functionalized graphene oxide/PVDF dual-layer flat sheet membrane for desalination via air gap membrane distillation. *Desalination*, 2018. **428**: pp. 227–239.
71. Anand, A. et al., Graphene-based nanofiltration membranes for improving salt rejection, water flux and antifouling—A review. *Desalination*, 2018. **429**: pp. 119–133.
72. Bhadra, M., S. Roy, and S. Mitra, Desalination across a graphene oxide membrane via direct contact membrane distillation. *Desalination*, 2016. **378**: pp. 37–43.
73. Tang, C. et al., Biomimetic aquaporin membranes coming of age. *Desalination*, 2015. **368**: pp. 89–105.
74. Li, X. et al., Nature gives the best solution for desalination: Aquaporin-based hollow fiber composite membrane with superior performance. *Journal of Membrane Science*, 2015. **494**: pp. 68–77.
75. Zou, T. et al., Fouling behavior and scaling mitigation strategy of CaSO 4 in submerged vacuum membrane distillation. *Desalination*, 2018. **425**: pp. 86–93.

76. Ko, Y. et al., Comparison of fouling behaviors of hydrophobic microporous membranes in pressure- and temperature-driven separation processes. *Desalination*, 2018. **428**: pp. 264–271.
77. Jung, Y. et al., Applications of nisin for biofouling mitigation of reverse osmosis membranes. *Desalination*, 2018. **429**: pp. 52–59.
78. Hou, D. et al., A novel dual-layer composite membrane with underwater-superoleophobic/hydrophobic asymmetric wettability for robust oil-fouling resistance in membrane distillation desalination. *Desalination*, 2018. **428**: pp. 240–249.
79. Goh, P.S. et al., Membrane fouling in desalination and its mitigation strategies. *Desalination*, 2018. **425**: pp. 130–155.
80. Sim, L.N. et al., A review of fouling indices and monitoring techniques for reverse osmosis. *Desalination*, 2017. **434**: pp. 169–188.
81. Kabir, E. et al., Solar energy: Potential and future prospects. *Renewable and Sustainable Energy Reviews*, 2018. **82**: pp. 894–900.
82. Kannan, N. and D. Vakeesan, Solar energy for future world: A review. *Renewable and Sustainable Energy Reviews*, 2016. **62**: pp. 1092–1105.
83. Compain, P., Solar energy for water desalination. *Procedia Engineering*, 2012. **46**: pp. 220–227.
84. Chandrashekara, M. and A. Yadav, Water desalination system using solar heat: A review. *Renewable and Sustainable Energy Reviews*, 2017. **67**: pp. 1308–1330.
85. Shalaby, S.M., Reverse osmosis desalination powered by photovoltaic and solar Rankine cycle power systems: A review. *Renewable and Sustainable Energy Reviews*, 2017. **73**: pp. 789–797.
86. United Nations, General Assembly, The human right to water and sanitation. 2010 ; Retrieved December 15, 2018 from http://www.un.org/en/ga/search/view_doc.asp?symbol=A/RES/64/292.
87. El-Nashar, A.M. and M. Samad, The solar desalination plant in Abu Dhabi: 13 years of performance and operation history. *Renewable Energy*, 1998. **14**(1): pp. 263–274.

2

Membrane Distillation Desalination Principles and Configurations

2.1 Membrane Distillation: Fast Growing Research Topic for Desalination

The first patent granted for the recovery of demineralized water from saline waters using membrane distillation was in 1967 [1]. The invention showed the potential for membrane distillation to produce very high purity water from saline solutions. The high purity of water produced means a wide range of application of membrane distillation can be envisioned. Unfortunately, this membrane separation technology did not succeed to establish itself as a viable commercial separation technique due to a number of factors, namely lack of thorough membrane characterization, cost of membrane fabrication, lack of data on membrane stability and mechanical strength. The first reported assessment of membrane distillation performance for desalination was reported by Hanbury and Hodgkiess [2]. They conducted experiments on membrane distillation using a polytetrafluorethylene (PTFE) membrane in a flat Perspex module. Synthetic brine (20,000 ppm NaCl) was used as hot feed. They measured the electrical conductivity of the distillate and found it of excellent quality. Moreover, they estimated distillation heat transfer coefficients assuming heat transfer analogy with multistage flash distillation. Their results showed a poor thermal performance, worse than published MSF thermal performance, hinting that a very large membrane surface area would be required to achieve results comparable to a once through MSF unit. They indicated that PTFE membranes would be costly and other "hydrophobic membrane" materials should be investigated, stating scaling was not investigated. This simple but informative set of results, alongside the patent by Svenska Utvecklings AB (SU, Sweden) described by Carlsson [3], the field tests reported by Andersson et al. [4] and the theoretical transport phenomena for membrane distillation put forward by Schofield et al. [5], had profound implications in future membrane distillation research directions a decade later.

Work on applications of membrane distillation (MD) for the treatment of aqueous solutions was among the early publications on MD [6,7] and continued to these days [8–11]. The promising potential of membrane distillation in a variety of commercial applications, including the growing lucrative seawater desalination market, led to a considerable growth in academic and industrial research. An extract from the Scopus database shown in Figure 2.1 depicts an exponential growth in the number of refereed papers on membrane distillation from 1985 to 2017 (with a snapshot in January 2018). In fact, the total number of papers in the Scopus database on membrane distillation in the period 1985 to January 2018 amounted to 1844 from which 785 were dedicated to "desalination," which is just over 42%. In addition, the Google patent database from 1979 to January 2018 indicates 650 patents were granted or published, showing the surge in commercial interest. Some of the reasons for this strong interest in academic research, especially for graduate research, include a synergistic combination of well understood and documented transport phenomena and experimental validation using relatively low-cost laboratory equipment and widely available computational tools. Research output on membrane distillation emanates from all continents, specially from water-stressed regions.

In this book, established principles of membrane distillation and recent developments will be reviewed and discussed, emphasizing the progress toward desalination prospects for commercial application and utilization in remote off-grid regions to supply freshwater to underprivileged communities.

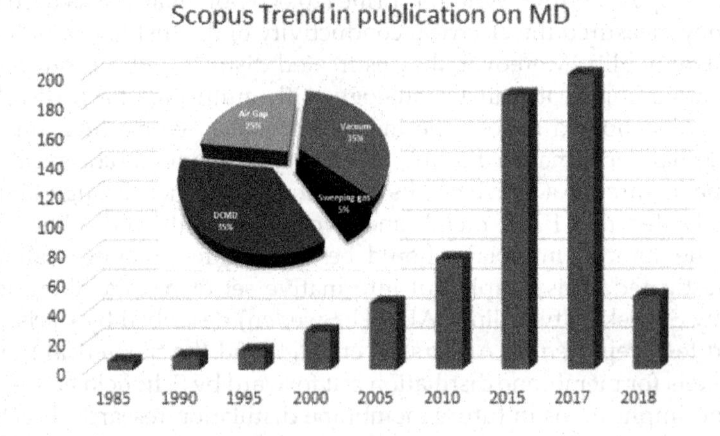

FIGURE 2.1
Growth in the number of refereed papers on membrane distillation from 1985 to 2017 (with a snapshot in January 2018).

2.2 Membrane Distillation Principles

The name "membrane distillation" for the thermally based membrane separation technique this book will cover for desalination was adopted in 1985 in Rome, Italy, by a working party made up of members representing universities from Italy, the Netherlands, Japan, Australia and a German company (Enka, that later became Membrana and was acquired by polypore in 2002 then by 3 M in 2015). The purpose of the meeting was to recognize membrane distillation as distinct from other apparently similar membrane techniques and put forward a terminology that became specific to membrane distillation.

The outcome of the membrane distillation nomenclature meeting yielded the following original characteristics [12]:

1. The membrane should be porous;
2. The membrane should not be wetted by process liquids;
3. No capillary condensation should take place inside the pores of the membranes;
4. Only vapor should be transported through the pores of the membrane;
5. The membrane must not alter the vapor equilibrium of the different components in the process liquids;
6. At least one side of the membrane should be in direct contact with the process liquid; and
7. For each component, the driving force of the membrane operation is a partial pressure gradient in the vapor phase.

Smolders and Franken [12] also reported that the literature described several "embodiments" in membrane distillation systems. These days we call these "embodiments" configurations and they are exactly how Smolders and Franken [12] described them in 1989:

- Direct contact membrane distillation
- Gas gap (air gap) membrane distillation
- Low pressure membrane distillation
- Sweeping gas membrane distillation

Each of these configurations will be discussed in the following sections of this chapter.

There are variants of the above four configurations reported in the literature: liquid gap membrane distillation (LGMD), also referred to as water gap membrane distillation (WGMD) [13,14], vacuum assisted air gap membrane distillation (VA-AGMD) [15], thermostatic sweeping gas membrane distillation (TSGMD) [16,17], which is a combination of SGMD and AGMD, and material gap membrane distillation (MGMD) [18]. These variants are less well used but show up occasionally in the literature for their unique features. They probably need more work to demonstrate their attributes.

In terms of operational features, the four main configurations have common and specific features: All require that the hot feed aqueous liquid be in contact with the microporous hydrophobic membrane where only vapor (of water for desalination, or vapor of any volatile solute present in water mixed with water vapor) is allowed to diffuse through the membrane pores to the cooler side. It is in the cooler side where the vapor is condensed at lower temperature either inside the membrane module or outside, that fabrication features make the configurations distinct from each other. The temperatures and pressure in the hot side are moderately low, typically 50°C–80°C and around atmospheric pressure. Temperatures lower than 50°C will not give significant vapor pressure at the liquid-membrane interface while temperatures higher than 80°C would require more energy input and would defeat the purpose of using low energy and low-grade heat, recalling that this is one of the attractive features of membrane distillation for desalination.

For systems that may have a pressure drop between the inlet and outlet of the hot side, slightly higher pressures may be used but, in any case, these should be lower than the membrane liquid entry pressure (LEP). This important point will be discussed further in the following chapter on membranes owing to its relevance to membrane wetting risks.

2.3 Direct Contact Membrane Distillation (DCMD)

Direct contact membrane distillation (DCMD) is one of the two most widely used configurations among the research community [19–47], as seen in Figure 2.1.

In DCMD both sides of the membrane are in contact with the liquids, one being the hot feed side, typically seawater or salt water and the other being the cold side acting as the condensing medium. The flow of hot and cold streams can be counter-current, as shown in Figure 2.2, or co-current [48,49]. The effect of flow direction on flux will be discussed in the chapter dedicated to module design. The widespread use of DCMD can be attributed to its relatively simple mechanical design, especially for flat sheet membranes. Figure 2.3 depicts a laboratory set-up for a DCMD [50] with full auxiliary items such as feed tanks, feed pumps, digital balances, temperature and pressure sensors and mass flow meters, all capable of displaying analog displays or logging digitally readings to a computer through appropriate data acquisition hardware-software. Figure 2.4 shows the actual DCMD

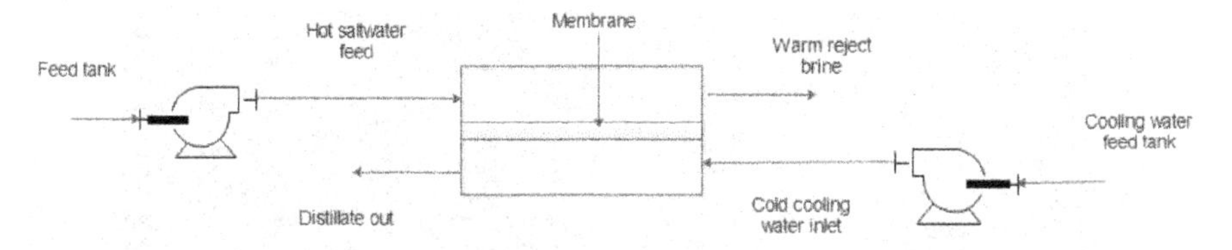

FIGURE 2.2
Counter-current flow DCMD system.

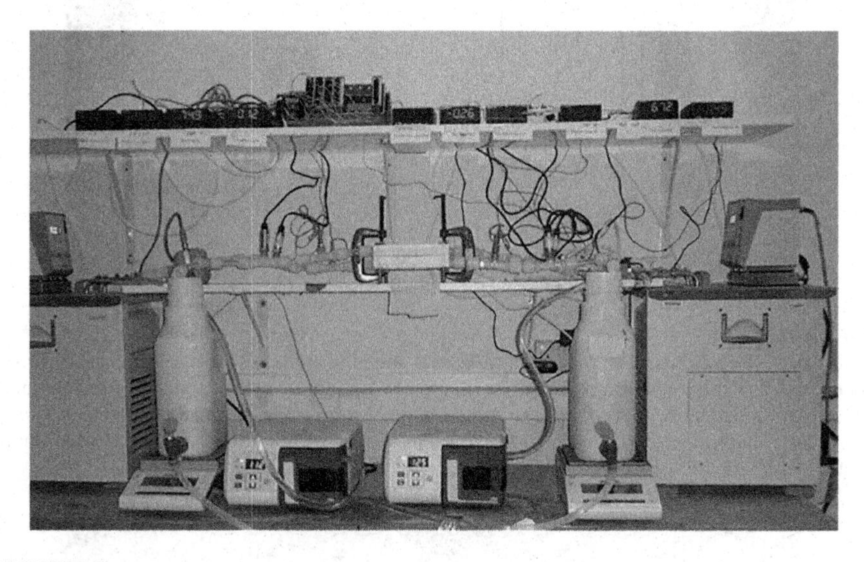

FIGURE 2.3
Laboratory set-up of DCMD system. (From Benyahia, F., Evaluation of membrane distillation to augment water production from thermal desalination brines by leveraging low grade waste heat, *Environmental Engineering Research Capacity Building Project*, ConocoPhillips (Global Water Sustainability Center), Doha Qatar (QUEX-CWT-11/12-5), Qatar University, 2017.)

module [50] (open) with clearly visible flow channels for the hot/cold sides and tools to cut to measure the flat sheet membrane. Figure 2.5 depicts an optional spacer [50] (mesh type). Like in the case of reverse osmosis, spacers have been used to promote flow turbulence in narrow channels. This is an important topic that will be discussed in detail in the chapters on module design and modeling.

The DCMD configuration can also accommodate hollow fiber membranes as bundles inside a cylindrical shell module. Figure 2.6 show (a) laboratory hollow fiber modules and (b) side end of a hollow fiber bundle [50].

Though DCMD configuration has been reported to give relatively high distillate fluxes, it suffers from low energy efficiency with heat losses through membrane conduction [51]. Operating at higher temperatures may minimize heat losses but not eliminate them.

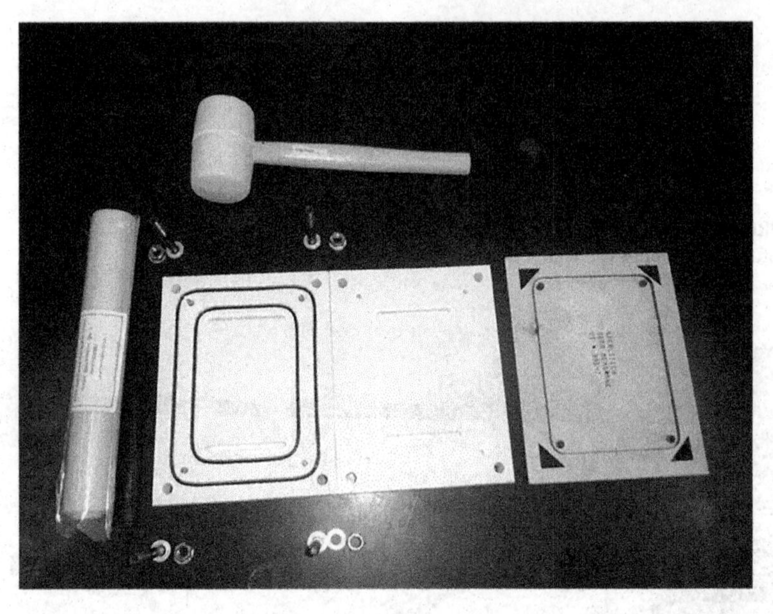

FIGURE 2.4
Laboratory membrane distillation module and membrane distillation cutting tool. (From Benyahia, F., Evaluation of membrane distillation to augment water production from thermal desalination brines by leveraging low grade waste heat, *Environmental Engineering Research Capacity Building Project*, ConocoPhillips (Global Water Sustainability Center), Doha Qatar (QUEX-CWT-11/12-5), Qatar University, 2017.)

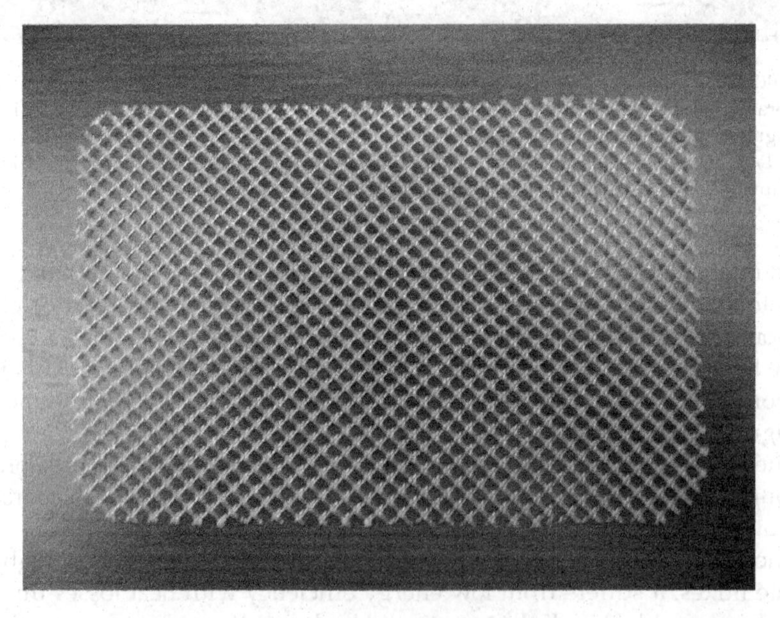

FIGURE 2.5
Membrane distillation module spacer.

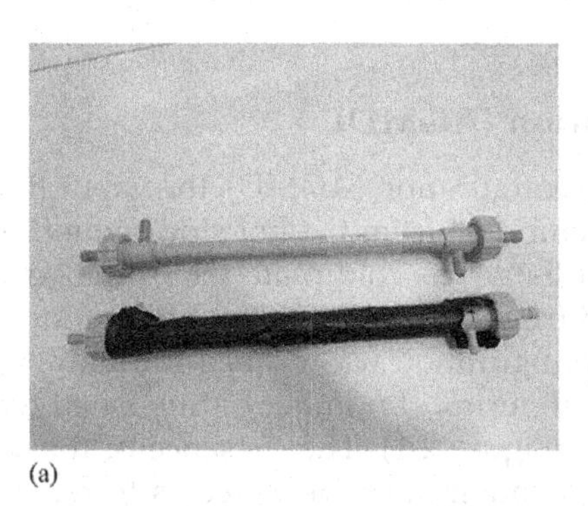

(a)

(b)

FIGURE 2.6
(a) Laboratory hollow fiber modules and (b) the side end of a hollow fiber bundle. (From Benyahia, F., Evaluation of membrane distillation to augment water production from thermal desalination brines by leveraging low grade waste heat, *Environmental Engineering Research Capacity Building Project*, ConocoPhillips (Global Water Sustainability Center), Doha Qatar (QUEX-CWT-11/12-5), Qatar University, 2017.)

2.4 Vacuum Membrane Distillation (VMD)

Vacuum membrane distillation (VMD) is now as popular as DCMD in terms popularity in research as evidenced by the fraction of papers published and shown in the Scopus database (Figure 2.1).

In the vacuum membrane distillation configuration, the permeate (distillate) side is under low pressure and the vapor that diffuses through the membrane is recovered and condensed outside the membrane module, as shown in Figure 2.7. Mechanically, the VMD system is more elaborate than DCMD and can be prone to membrane wetting [52]. However, it has been reported to have a better energy efficiency and high flux [52]. VMD was initially intended for water treatment to remove volatile solutes in water [53]. With the emergence of gas-to-liquids (GTL) technology to produce clean fuels from natural gas, GTL wastewater that contains substantial amounts of alcohols can be treated with VMD to recover these alcohols. Recently, a pilot plant using the Memsys multistage vacuum membrane distillation has shown that VMD can be used effectively in desalinating high salinity Arabian Gulf seawater [54].

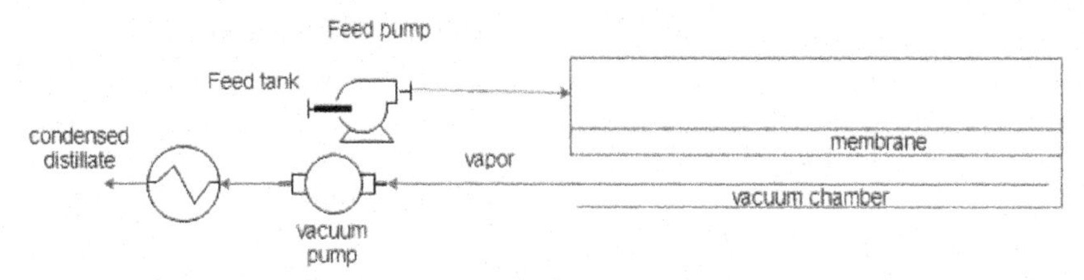

FIGURE 2.7
Schematic representation of a vacuum membrane distillation configuration.

2.5 Air Gap Membrane Distillation (AGMD)

In the air gap membrane distillation configuration (AGMD), the vapor that diffuses through the hydrophobic membrane from the hot side condenses on a cold plate and recovered as distillate. The cold plate requires a coolant to evacuate the latent heat gained from the vapor. This configuration is depicted in Figure 2.8. The SGMD configuration was ranked third among the four configurations in popularity in terms of number of published refereed papers in the Scopus database (Figure 2.1). The presence of the air gap between the membrane and the condensation plate seems to reduce significantly the heat losses [55] mentioned previously for the DCMD, and promises a good flux. As a result, there is an increasing interest in AGMD [56–77].

2.6 Sweeping Gas Membrane Distillation (SGMD)

The sweeping gas membrane distillation (SGMD) received considerably less attention than the other three membrane distillation configurations, as evidenced by the number of publications on membrane distillation in the Scopus database (Figure 2.1). In the SGMD configuration, a carrier gas called sweeping gas pushes the vapor toward an external condenser where distillate is recovered. This configuration is depicted in Figure 2.9.

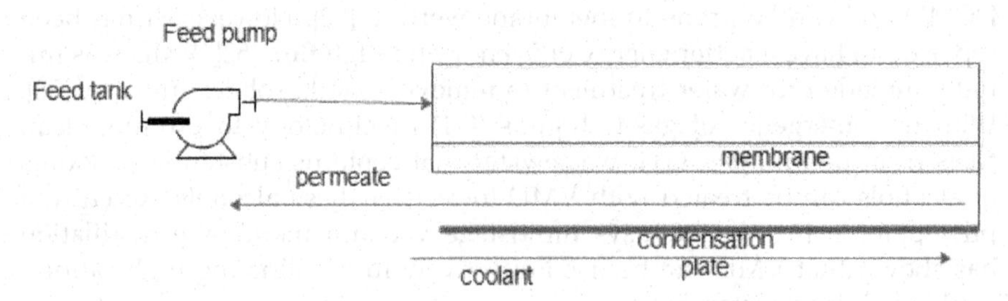

FIGURE 2.8
Schematic representation of an air gap membrane distillation configuration.

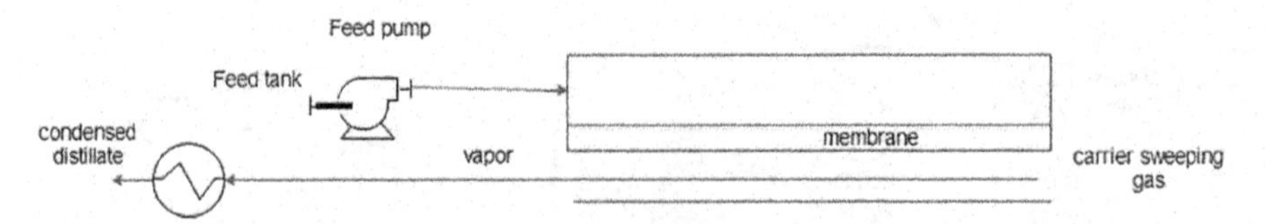

FIGURE 2.9
Schematic representation of a sweeping gas membrane distillation configuration.

Nevertheless, a number of publications were contributed to using this configuration that is reported to be suitable for water treatment to remove volatile solutes in aqueous streams or to concentrate solutions by evaporating water, but also to desalinate salt water [16,17,78–90].

2.7 Concluding Remarks

Membrane distillation (MD) is now an established field for research in separation processes with excellent prospects to desalinate seawater to produce good quality freshwater. At the heart of the membrane distillation process is the hydrophobic microporous membrane. It is the very essence of selectivity toward pure distillate.

The environment in which the hydrophobic membrane operates is influenced by the configuration, thus affecting the flux that is a measure of productivity.

Since the MD process is governed by the simultaneous transport of heat and mass of the species involved, it has been the subject of intense scrutiny from the modeling angle. This topic deserves a full chapter owing to the recent developments and improvements in matching model predictions with observed experimental results. In particular, the transport properties deserve a renewed scrutiny to review assumptions and other considerations such as turbulence in the modules.

The operational problems and proposed solutions for the mature reverse osmosis technique gave us ideas on how to move forward with membrane distillation and make significant progress to enable MD deployment where it is most needed: in regions that suffer from water stress leveraging the availability of renewable energy like solar energy but also thermally integrating MD in chemical processes where a vast amount of low-grade heat is currently dissipated to the environment.

Some of the outstanding challenges in MD are quite similar to reverse osmosis: fouling and membrane stability.

All in all, the solutions must lead to affordable solutions for the underprivileged communities.

References

1. Weyl, P.K., *Recovery of Demineralized Water from Saline Waters*. 1967, Research Corporation.
2. Hanbury, W.T. and T. Hodgkiess, Membrane distillation—An assessment. *Desalination*, 1985. **56**: pp. 287–297.

3. Carlsson, L., The new generation in sea-water desalination Su membrane distillation system. *Desalination*, 1983. **45**: pp. 221–222.
4. Andersson, S.I., N. Kjellander, and B. Rodesjö, Design and field tests of a new membrane distillation desalination process. *Desalination*, 1985. **56**: pp. 345–354.
5. Schofield, R.W., A.G. Fane, and C.J.D. Fell, Heat and mass-transfer in membrane distillation. *Journal of Membrane Science*, 1987. **33**(3): pp. 299–313.
6. Drioli, E., Y.L. Wu, and V. Calabro, Membrane distillation in the treatment of aqueous-solutions. *Journal of Membrane Science*, 1987. **33**(3): pp. 277–284.
7. Tomaszewska, M., Membrane distillation—Examples of applications in technology and environmental protection. *Polish Journal of Environmental Studies*, 2000. **9**(1): pp. 27–36.
8. Kiai, H. et al., Application of membrane distillation technology in the treatment of table olive wastewaters for phenolic compounds concentration and high quality water production. *Chemical Engineering and Processing: Process Intensification*, 2014. **86**: pp. 153–161.
9. El-Abbassi, A. et al., Treatment of olive mill wastewater by membrane distillation using polytetrafluoroethylene membranes. *Separation and Purification Technology*, 2012. **98**: pp. 55–61.
10. Minier-Matar, J. et al., Application of membrane contactors to remove hydrogen sulfide from sour water. *Journal of Membrane Science*, 2017. **541**: pp. 378–385.
11. Quist-Jensen, C.A. et al., Direct contact membrane distillation for the concentration of clarified orange juice. *Journal of Food Engineering*, 2016. **187**: pp. 37–43.
12. Smolders, K. and A.C.M. Franken, Terminology for membrane distillation. *Desalination*, 1989. **72**(3): pp. 249–262.
13. Ugrozov, V.V. and L.I. Kataeva, Mathematical modeling of membrane distiller with liquid gap. *Desalination*, 2004. **168**: pp. 347–353.
14. Khalifa, A.E. and S.M. Alawad, Air gap and water gap multistage membrane distillation for water desalination. *Desalination*, 2018. **437**: pp. 175–183.
15. Prince, J.A., G. Singh, and T.S. Shanmugasundaram, A vacuum air gap membrane distillation system for desalination. Patent WO2013/151498Al. 2013.
16. Rivier, C.A. et al., Separation of binary mixtures by thermostatic sweeping gas membrane distillation I: Theory and simulations. *Journal of Membrane Science*, 2002. **201**(1–2): pp. 1–16.
17. Garcia-Payo, M.C. et al., Separation of binary mixtures by thermostatic sweeping gas membrane distillation—II. Experimental results with aqueous formic acid solutions. *Journal of Membrane Science*, 2002. **198**(2): pp. 197–210.
18. Francis, L. et al., Material gap membrane distillation: A new design for water vapor flux enhancement. *Journal of Membrane Science*, 2013. **448**: pp. 240–247.
19. Lee, J.-G. et al., Influence of high range of mass transfer coefficient and convection heat transfer on direct contact membrane distillation performance. *Desalination*, 2018. **426**: pp. 127–134.
20. Ali, A. et al., Designing and optimization of continuous direct contact membrane distillation process. *Desalination*, 2018. **426**: pp. 97–107.
21. Soukane, S. et al., Effect of feed flow pattern on the distribution of permeate fluxes in desalination by direct contact membrane distillation. *Desalination*, 2017. **418**: pp. 43–59.

22. Seo, J., Y.M. Kim, and J.H. Kim, Spacer optimization strategy for direct contact membrane distillation: Shapes, configurations, diameters, and numbers of spacer filaments. *Desalination*, 2017. **417**: pp. 9–18.
23. Kurdian, A.R. et al., Modeling of direct contact membrane distillation process: Flux prediction of sodium sulfate and sodium chloride solutions. *Desalination*, 2013. **323**: pp. 75–82.
24. Koo, J. et al., Experimental comparison of direct contact membrane distillation (DCMD) with vacuum membrane distillation (VMD). *Desalination and Water Treatment*, 2013. **51**(31–33): pp. 6299–6309.
25. Ibrahim, S.S. and Q.F. Alsalhy, Modeling and simulation for direct contact membrane distillation in hollow fiber modules. *AICHE Journal*, 2013. **59**(2): pp. 589–603.
26. Ho, C.D., T.J. Yang, and Y.C. Chuang, Performance improvement of counter-current-flow direct contact membrane distillation in seawater desalination systems. *Desalination and Water Treatment*, 2013. **51**(25–27): pp. 5113–5120.
27. He, F., J. Gilron, and K.K. Sirkar, High water recovery in direct contact membrane distillation using a series of cascades. *Desalination*, 2013. **323**: pp. 48–54.
28. Dumée, L.F. et al., The role of membrane surface energy on direct contact membrane distillation performance. *Desalination*, 2013. **323**: pp. 22–30.
29. Shirazi, M.M.A., A. Kargari, and M.J.A. Shirazi, Direct contact membrane distillation for seawater desalination. *Desalination and Water Treatment*, 2012. **49**(1–3): pp. 368–375.
30. Zuo, G. et al., Energy efficiency evaluation and economic analyses of direct contact membrane distillation system using Aspen Plus. *Desalination*, 2011. **283**: pp. 237–244.
31. Hwang, H.J. et al., Direct contact membrane distillation (DCMD): Experimental study on the commercial PTFE membrane and modeling. *Journal of Membrane Science*, 2011. **371**(1–2): pp. 90–98.
32. Chen, T.-C. and C.-D. Ho, Immediate assisted solar direct contact membrane distillation in saline water desalination. *Journal of Membrane Science*, 2010. **358**(1): pp. 122–130.
33. Bui, V.A., L.T.T. Vu, and M.H. Nguyen, Simulation and optimisation of direct contact membrane distillation for energy efficiency. *Desalination*, 2010. **259**(1–3): pp. 29–37.
34. Chen, T.-C., C.-D. Ho, and H.-M. Yeh, Theoretical modeling and experimental analysis of direct contact membrane distillation. *Journal of Membrane Science*, 2009. **330**(1–2): pp. 279–287.
35. Song, L. et al., Direct contact membrane distillation-based desalination: Novel membranes, devices, larger-scale studies, and a model. *Industrial & Engineering Chemistry Research*, 2007. **46**(8): pp. 2307–2323.
36. Martínez, L. and J.M. Rodríguez-Maroto, Effects of membrane and module design improvements on flux in direct contact membrane distillation. *Desalination*, 2007. **205**(1–3): pp. 97–103.
37. Martinez, L. and J.M. Rodriguez-Maroto, On transport resistances in direct contact membrane distillation. *Journal of Membrane Science*, 2007. **295**(1–2): pp. 28–39.
38. Khayet, M. et al., Design of novel direct contact membrane distillation membranes. *Desalination*, 2006. **192**(1–3): pp. 105–111.

39. Termpiyakul, P., R. Jiraratananon, and S. Srisurichan, Heat and mass transfer characteristics of a direct contact membrane distillation process for desalination. *Desalination*, 2005. **177**(1–3): pp. 133–141.
40. Li, B. and K.K. Sirkar, Novel membrane and device for direct contact membrane distillation-based desalination process. *Industrial & Engineering Chemistry Research*, 2004. **43**(17): pp. 5300–5309.
41. Cath, T.Y., V.D. Adams, and A.E. Childress, Experimental study of desalination using direct contact membrane distillation: A new approach to flux enhancement. *Journal of Membrane Science*, 2004. **228**(1): pp. 5–16.
42. Phattaranawik, J., Effect of pore size distribution and air flux on mass transport in direct contact membrane distillation. *Journal of Membrane Science*, 2003. **215**(1–2): pp. 75–85.
43. Hsu, S.T., K.T. Cheng, and J.S. Chiou, Seawater desalination by direct contact membrane distillation. *Desalination*, 2002. **143**(3): pp. 279–287.
44. Phattaranawik, J. et al., Mass flux enhancement using spacer filled channels in direct contact membrane distillation. *Journal of Membrane Science*, 2001. **187**(1): pp. 193–201.
45. Phattaranawik, J. and R. Jiraratananon, Direct contact membrane distillation: Effect of mass transfer on heat transfer. *Journal of Membrane Science*, 2001. **188**(1): pp. 137–143.
46. Laganà, F., G. Barbieri, and E. Drioli, Direct contact membrane distillation: Modelling and concentration experiments. *Journal of Membrane Science*, 2000. **166**(1): pp. 1–11.
47. Lawson, K.W. and D.R. Lloyd, Membrane distillation: II. Direct contact MD. *Journal of Membrane Science*, 1996. **120**(1): pp. 123–133.
48. Ho, C.D. et al., Performance improvement on distillate flux of countercurrent-flow direct contact membrane distillation systems. *Desalination*, 2014. **338**: pp. 26–32.
49. Cheng, L.-H. et al., Spatial variations of DCMD performance for desalination through countercurrent hollow fiber modules. *Desalination*, 2008. **234**(1–3): pp. 323–334.
50. Benyahia, F., Evaluation of membrane distillation to augment water production from thermal desalination brines by leveraging low grade waste heat. *Environmental Engineering Research Capacity Building Project*. 2017, ConocoPhillips (Global Water Sustainablity Center), Doha Qatar (QUEX-CWT-11/12-5): Qatar University.
51. Chernyshov, M.N., G.W. Meindersma, and A.B. de Haan, Modelling temperature and salt concentration distribution in membrane distillation feed channel. *Desalination*, 2003. **157**(1): pp. 315–324.
52. Abu-Zeid, M.A.E.-R. et al., A comprehensive review of vacuum membrane distillation technique. *Desalination*, 2015. **356**: pp. 1–14.
53. Bandini, S., C. Gostoli, and G.C. Sarti, Separation efficiency in vacuum membrane distillation. *Journal of Membrane Science*, 1992. **73**(2–3): pp. 217–229.
54. Minier-Matar, J. et al., Field evaluation of membrane distillation technologies for desalination of highly saline brines. *Desalination*, 2014. **351**: pp. 101–108.
55. Jönsson, A.S., R. Wimmerstedt, and A.C. Harrysson, Membrane distillation—A theoretical study of evaporation through microporous membranes. *Desalination*, 1985. **56**: pp. 237–249.

56. Bindels, M., N. Brand, and B. Nelemans, Modeling of semibatch air gap membrane distillation. *Desalination*, 2018. **430**: pp. 98–106.
57. Khalifa, A.E., S.M. Alawad, and M.A. Antar, Parallel and series multistage air gap membrane distillation. *Desalination*, 2017. **417**: pp. 69–76.
58. Janajreh, I. et al., Numerical investigation of air gap membrane distillation (AGMD): Seeking optimal performance. *Desalination*, 2017. **424**: pp. 122–130.
59. Hitsov, I. et al., Full-scale validated air gap membrane distillation (AGMD) model without calibration parameters. *Journal of Membrane Science*, 2017. **533**: pp. 309–320.
60. Alkhudhiri, A. and N. Hilal, Air gap membrane distillation: A detailed study of high saline solution. *Desalination*, 2017. **403**: pp. 179–186.
61. Aryapratama, R. et al., Performance evaluation of hollow fiber air gap membrane distillation module with multiple cooling channels. *Desalination*, 2016. **385**: pp. 58–68.
62. He, Q.F. et al., Modeling and optimization of air gap membrane distillation system for desalination. *Desalination*, 2014. **354**: pp. 68–75.
63. Summers, E.K. and J.H. Lienhard, Experimental study of thermal performance in air gap membrane distillation systems, including the direct solar heating of membranes. *Desalination*, 2013. **330**: pp. 100–111.
64. Summers, E.K. and J.H. Lienhard, A novel solar-driven air gap membrane distillation system. *Desalination and Water Treatment*, 2013. **51**(7–9): pp. 1344–1351.
65. Alkhudhiri, A., N. Darwish, and N. Hilal, Treatment of saline solutions using air gap membrane distillation: Experimental study. *Desalination*, 2013. **323**: pp. 2–7.
66. Singh, D. and K.K. Sirkar, Desalination by air gap membrane distillation using a two hollow-fiber-set membrane module. *Journal of Membrane Science*, 2012. *421–422*: pp. 172–179.
67. Khayet, M. and C. Cojocaru, Air gap membrane distillation: Desalination, modeling and optimization. *Desalination*, 2012. **287**: pp. 138–145.
68. Alkhudhiri, A., N. Darwish, and N. Hilal, Treatment of high salinity solutions: Application of air gap membrane distillation. *Desalination*, 2012. **287**: pp. 55–60.
69. Guillén-Burrieza, E. et al., Experimental analysis of an air gap membrane distillation solar desalination pilot system. *Journal of Membrane Science*, 2011. **379**(1–2): pp. 386–396.
70. Matheswaran, M. et al., Factors affecting flux and water separation performance in air gap membrane distillation. *Journal of Industrial and Engineering Chemistry*, 2007. **13**(6): pp. 965–970.
71. Meindersma, G.W., C.M. Guijt, and A.B. de Haan, Water recycling and desalination by air gap membrane distillation. *Environmental Progress*, 2005. **24**(4): pp. 434–441.
72. Guijt, C. et al., Air gap membrane distillation 2: Model validation and hollow fibre module performance analysis. *Separation and Purification Technology*, 2005. **43**(3): pp. 245–255.
73. Guijt, C. et al., Air gap membrane distillation1: Modelling and mass transport properties for hollow fibre membranes. *Separation and Purification Technology*, 2005. **43**(3): pp. 233–244.
74. El Amali, A., S. Bouguecha, and M. Maalej, Experimental study of air gap and direct contact membrane distillation configurations: Application to geothermal and seawater desalination. *Desalination*, 2004. **168**: pp. 357–357.

75. Abu Al-Rub, F.A., F. Banat, and K. Bani-Melhem, Sensitivity analysis of air gap membrane distillation. *Separation Science and Technology*, 2003. **38**(15): pp. 3645–3667.
76. Izquierdo-Gil, M.A., M.C. Garcia-Payo, and C. Fernandez-Pineda, Air gap membrane distillation of sucrose aqueous solutions. *Journal of Membrane Science*, 1999. **155**(2): pp. 291–307.
77. Liu, G.L. et al., Theoretical and experimental studies on air gap membrane distillation. *Heat and Mass Transfer*, 1998. **34**(4): pp. 329–335.
78. Huang, S.-M. et al., Sweeping air membrane distillation: Conjugate heat and mass transfer in a hollow fiber membrane tube bank with an in-line arrangement. *International Journal of Heat and Mass Transfer*, 2017. **108**: pp. 2191–2197.
79. Zhao, S.F. et al., Condensation, re-evaporation and associated heat transfer in membrane evaporation and sweeping gas membrane distillation. *Journal of Membrane Science*, 2015. **475**: pp. 445–454.
80. Shirazi, M.M.A., A. Kargari, and M. Tabatabaei, Sweeping gas membrane distillation (SGMD) as an alternative for integration of bioethanol processing: Study on a commercial membrane and operating parameters. *Chemical Engineering Communications*, 2015. **202**(4): pp. 457–466.
81. Zhao, S.F. et al., Condensation studies in membrane evaporation and sweeping gas membrane distillation. *Journal of Membrane Science*, 2014. **462**: pp. 9–16.
82. Shirazi, M.M.A. et al., Concentration of glycerol from dilute glycerol wastewater using sweeping gas membrane distillation. *Chemical Engineering and Processing*, 2014. **78**: pp. 58–66.
83. Khayet, M. and C. Cojocaru, Artificial neural network model for desalination by sweeping gas membrane distillation. *Desalination*, 2013. **308**: pp. 102–110.
84. Cojocaru, C. and M. Khayet, Sweeping gas membrane distillation of sucrose aqueous solutions: Response surface modeling and optimization. *Separation and Purification Technology*, 2011. **81**(1): pp. 12–24.
85. Charfi, K., M. Khayet, and M.J. Safi, Numerical simulation and experimental studies on heat and mass transfer using sweeping gas membrane distillation. *Desalination*, 2010. **259**(1–3): pp. 84–96.
86. Khayet, M., M.P. Godino, and J.I. Mengual, Theoretical and experimental studies on desalination using the sweeping gas membrane distillation method. *Desalination*, 2003. **157**(1–3): pp. 297–305.
87. Khayet, M., M.P. Godino, and J.I. Mengual, Thermal boundary layers in sweeping gas membrane distillation processes. *Aiche Journal*, 2002. **48**(7): pp. 1488–1497.
88. Khayet, M., P. Godino, and J.I. Mengual, Theory and experiments on sweeping gas membrane distillation. *Journal of Membrane Science*, 2000. **165**(2): pp. 261–272.
89. Khayet, M., P. Godino, and J.I. Mengual, Nature of flow on sweeping gas membrane distillation. *Journal of Membrane Science*, 2000. **170**(2): pp. 243–255.
90. Basini, L. et al., A desalination process through sweeping gas membrane distillation. *Desalination*, 1987. **64**: pp. 245–257.

3

Membranes for Membrane Distillation in Desalination

3.1 Introduction

In Chapter 2 some distinctive and important characteristics of membrane distillation were introduced [1]. Some of these are related to the membrane properties:

1. The membrane should be porous;
2. The membrane should not be wetted by process liquids.

Figure 3.1 depicts a diagram representing some of the physics underlying membrane distillation desalination incorporating the above properties. In Figure 3.1 only the water vapor from the hot side is allowed to penetrate the membrane pores and eventually condenses in the cold side for direct contact membrane configuration (DCMD) or is carried away to an external condenser for the other configurations. In the case of the DCMD, the interfaces between the hot side and cold side fluids and the hydrophobic membrane are extremely important since the phenomena that occur there affect to a great extent the overall performance of membrane distillation. That interface also includes the well-known boundary layer (on both hot and cold sides) of fluids where the stagnant film of fluid adds an additional resistance to heat and mass transport between the hot and cold side. In Figure 3.1, T_F and T_D are the bulk hot feed and bulk cold distillate temperatures, and the curve between them depicts the temperature profile that often represents the so-called temperature polarization effect studied in modeling work. It is important to note that interfacial temperatures cannot be easily measured and the easiest way to estimate them is through mathematical modeling of membrane distillation.

The second curve in Figure 3.1 on the hot feed side represents the change in salt water concentration between the bulk solution and the interface between the membrane and the liquid. One can note a slight increase in salinity (concentration of the species of interest, namely salt for desalination).

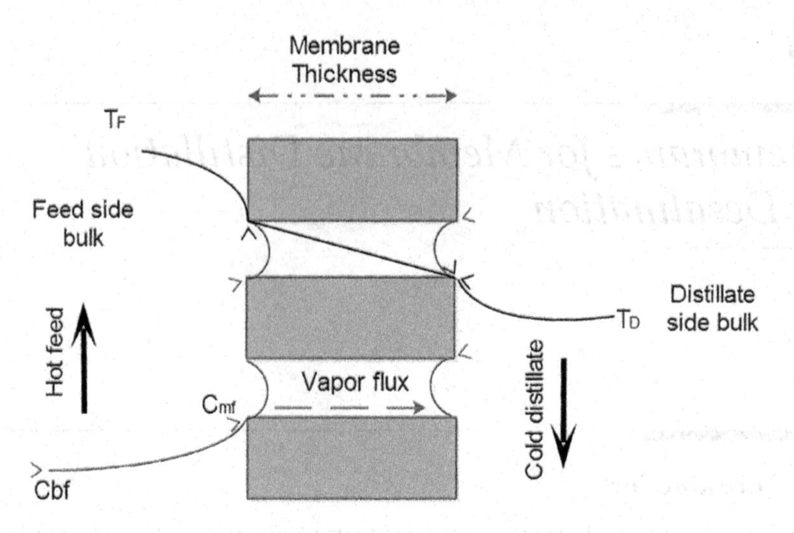

FIGURE 3.1
Schematic diagram of temperature and concentration polarization phenomena on the surface of a membrane distillation hydrophobic membrane.

C_{bf} represents the bulk solution feed concentration and C_{mf} represents the membrane surface concentration of the feed solution. Just like the case of temperature polarization, the concentration curve represents concentration polarization since the boundary layer adjacent to the membrane on the hot feed side loses some water vapor, and this naturally leads to a locally increased salt concentration. This phenomenon tends to occur all along the path of flow next to the membrane and therefore the values of C_{mf} will change accordingly. It is not easy to measure it, but it can be estimated computationally through modeling. A study of concentration polarization is critical in in-depth studies of membrane scaling mechanisms. In addition, temperature polarization and concentration polarization are inter-related and both affect the productivity of the membrane distillation system, the temperature effect being more important. This will be covered in detail in the chapter on modeling concepts in membrane distillation.

The negation of membrane wetting implies that the membrane should be hydrophobic while the porous nature indicates a morphology where microfiltration pore size range dominates. This chapter will be dedicated to a review of aspects of hydrophobic membrane preparation, characterization and shaping for ultimate utilization for desalination. A selection of important methods and techniques involved in the preparation and characterization of hydrophobic membranes for membrane distillation application will also be covered. Readers will be directed to more specialized sources for an in-depth coverage of techniques that are beyond the scope of this manuscript.

The historical lessons derived from the successful development of thermal and reverse osmosis desalination covered in Chapter 1 are immensely useful even if the mechanism of desalination and nature of membranes are different. These lessons will be invoked where appropriate in order to be objective with the scientific limits nature imposes and recognize the importance of mitigating practical problems that may lead to a deterioration in membrane distillation performance, namely flux decline and membrane wetting (poor quality distillate).

The remainder of this important chapter will explore important properties of membrane distillation membranes.

3.2 Membrane Hydrophobicity

Hydrophobicity of membranes in membrane distillation is usually determined experimentally using relatively simple methods such as the sessile drop technique [2,3]. However, underneath such advanced tools lay complex phenomena and surface thermodynamics that have been the subject of numerous studies for the past century and still going on for a wide variety of applications, including behavior in space [2–6]. The sessile drop technique is suitable for flat sheet membranes. For hollow fibers and other irregular shaped material in general, the Wilhelmy balance method can be used [3,7,8].

3.2.1 Definition of Hydrophobicity in Membranes for Membrane Distillation

A polymeric membrane is referred to as hydrophobic in the literature if the contact angle as measured (say by the sessile drop technique or equivalent depending on the membrane shape) is greater than 90. Conversely, if the contact angle is much less than 90 then the membrane is referred to as hydrophilic. Figure 3.2 depicts the classic diagram for describing the extent of hydrophobicity of membrane surfaces. The angle θ shown in Figure 3.1 is also described as Young's contact angle and can be found in Young's equation [3]:

$$\gamma_{lv} \cos\theta = \gamma_{sv} - \gamma_{sl} \tag{3.1}$$

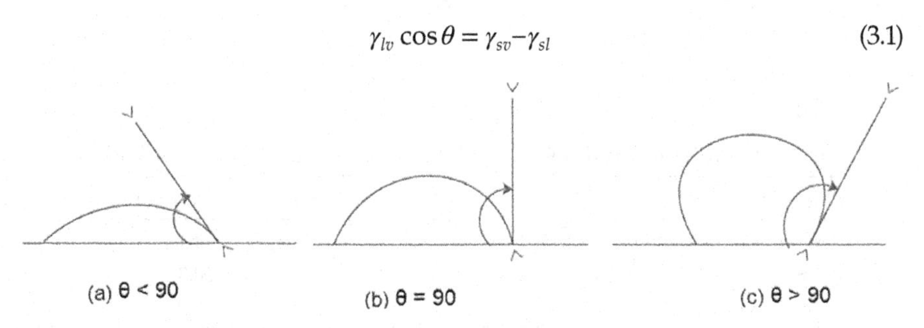

FIGURE 3.2
Extent of hydrophobicity of membrane surfaces as measured through contact angles.

where γ_{lv}, γ_{sv} and γ_{sl} represent the liquid vapor, solid vapor and solid liquid interfacial surface tensions respectively. Alongside this arbitrary definition of hydrophobicity, one can also say that a membrane is completely wetted if the contact angle θ is zero. If the contact angle is significantly greater than 90 (say 150 and over), the membrane is often referred to as "superhydrophobic."

3.3 Materials for Hydrophobic Membranes

The material of choice for membrane distillation are those polymers that are intrinsically hydrophobic (low surface energy) and can be processed into membranes (sheets or otherwise) with pore sizes within the microfiltration range (0.1–10 μm). Typically, polytetraethylene fluoride (PTFE), polyvinyledenfluoride (PVDF) and polypropylene (PP) are indeed used often in membrane distillation. Table 3.1 shows the surface free energy for PTFE, PVDF and PP.

The preparation of hydrophobic membranes for membrane distillation may involve a number of techniques that will provide the necessary pore size range (microfiltration range). Such techniques include sintering, stretching, track etching and phase inversion [9]. For instance, PTFE membranes can be made using sintering or stretching [10] while PVDF can be made using phase inversion [11,12]. Polypropylene (PP) can be made melt-extrusion followed by stretching [13] or using a phase inversion method. It is noted that isotactic PP that is highly crystalline is used for membrane distillation. All three polymers (PTFE, PVDF and PP) have excellent chemical and thermal stability properties in addition to non-wetting properties, and are therefore often the polymers of choice for membrane distillation.

Full details of sintering, stretching, track etching and phase inversion can be found elsewhere [9,14].

TABLE 3.1

Surface Energies of Commonly Used Polymers in Membrane Distillation

Polymer	Surface Energy (10^3 N/m)
Polytetrafluoroethylene (PTFE)	19.1
Polyvinylidenefluoride (PVDF)	30.3
Polypropylene (PP)	30.0

Source: Mulder, M., *Basic Principles of Membrane Technology*, Kluwer Academic Publishers, Dordrecht, the Netherlands, 1997.

3.4 Membrane Shape

In membrane distillation, the hydrophobic membranes are installed inside a module for the separation of freshwater from salty feed. There are two broad geometries of modules: rectangular, using flat sheet membranes or cylindrical (also known as tubular) using hollow fibers/capillaries as bundles in a shell and tube fashion. The cylindrical module can also accommodate spiral wound membranes [15,16] but these are indeed rare in membrane distillation publications.

Flat sheet membranes can be purchased as rolls or fabricated as large sheets and cut to size required for the dimensions of the housing modules.

Figure 3.3 shows laboratory flat sheets made of polypropylene and PTFE (apparently undistinguishable with the naked eye). Figure 3.4 shows an SEM image of a flat sheet polypropylene membrane. Figure 3.5 shows an SEM image of a flat sheet PTFE membrane.

Flat sheet membranes are traditionally prepared using well established techniques such as sintering, track-etching and phase inversion [9]. However, in recent years electrospinning became more widespread [17–27].

Hollow fiber membranes for membrane distillation usually have a diameter less than 0.5 mm. Their fabrication processes is different from those of flat sheets and comprise a variety of so-called spinning methods [9]:

- Dry spinning
- Wet spinning
- Melt spinning
- Electrospining [17,19–21,23–28]

FIGURE 3.3
Laboratory flat sheets made of polypropylene and PTFE for membrane distillation.

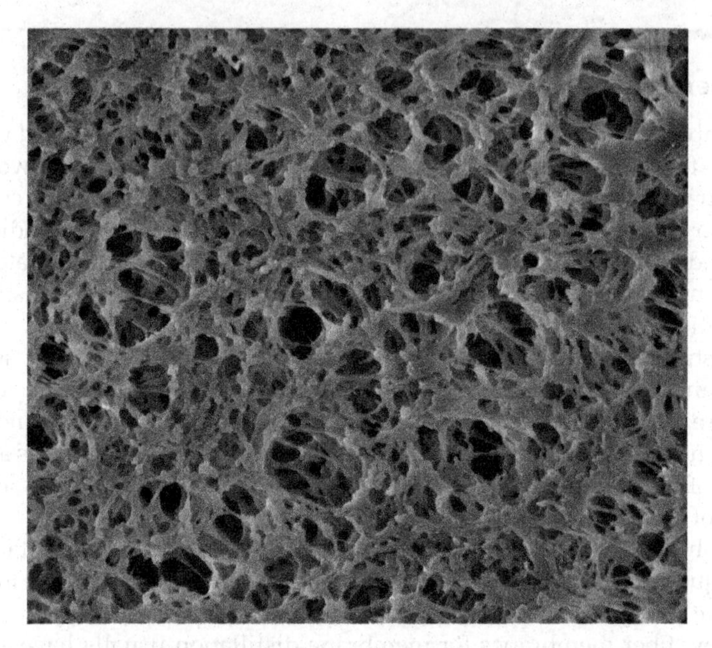

FIGURE 3.4
SEM image of a flat sheet polypropylene membrane for membrane distillation.

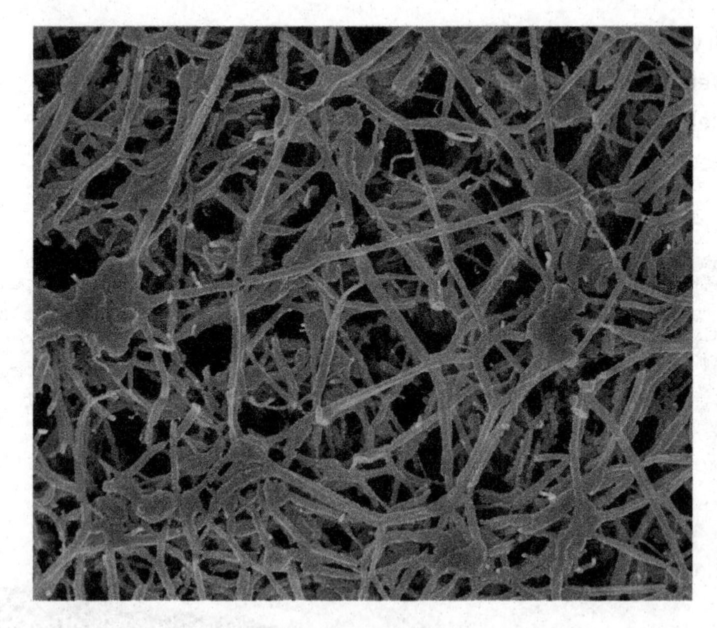

FIGURE 3.5
SEM image of a flat sheet PTFE membrane for membrane distillation.

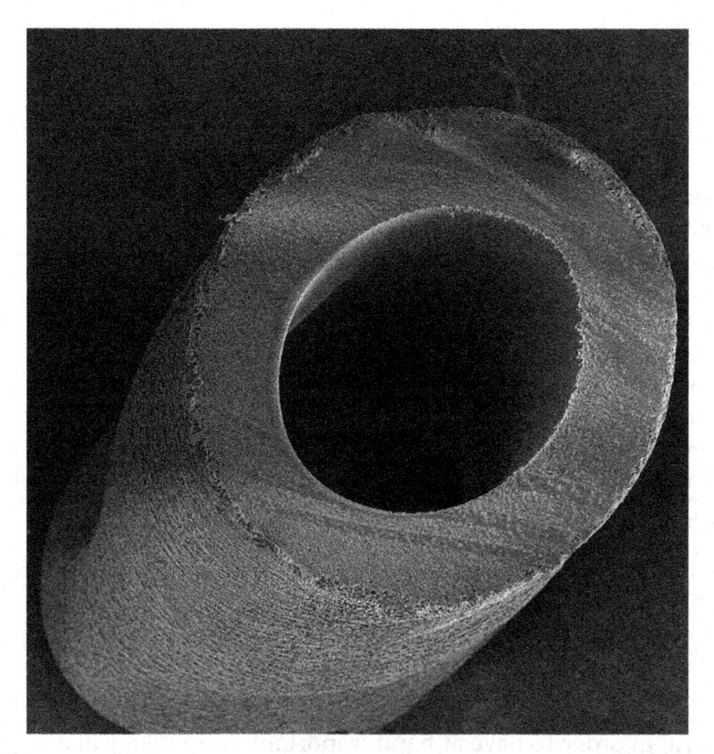

FIGURE 3.6
A section of a single hollow fiber membrane made from polypropylene for membrane distillation.

The spinning parameters affect the hollow fiber morphology and ultimately its performance in membrane distillation desalination [29,30].

Figure 3.6 shows a section of a single hollow fiber membrane made from polypropylene.

In practical applications, a number of hollow fiber membranes are housed in a tubular (cylindrical) module. This is often referred to as a "bundle" of tubular membranes.

The effect of membrane shape on performance will be covered in a separate chapter. Indeed, the performance is related to both membrane shape and module design and associated transfer and transport phenomena therein.

3.5 Hydrophobic Membrane Characterization

The ultimate flux permeance and associated efficiencies and productivity are related to the membrane properties as much as the operating condition. It is therefore important to characterize the hydrophobic membranes prior

TABLE 3.2

Hydrophobic Flat Sheet Membrane Properties and Their Impact on Performance

Properties	Performance Affected	Range of Properties Reported
Material	Surface energy	19–42 (× 10^3) N.m
Contact angle	Wetting	90–135[b]
Liquid entry pressure (LEP)	Wetting	0.13–4.6 bar
Nominal pore size	Flux[a]	0.02–1 μm
Porosity	Flux[a], strength	38%–90%
Thickness	Flux[a], strength	25–178 μm
Tortuosity	Flux[a]	1–4
Thermal conductivity	Flux[a]	0.03–0.06 W m^{-1} K^{-1}
Zeta potential		[c]
Tensile strength	Strength	3–60 MPa
Shape (flat sheet—hollow fiber)	Flux[a]	Higher for flat sheets

[a] Also energy efficiency.
[b] Higher values for modified membranes (superhydrophobic).
[c] Not available for unmodified membranes. This property in MD is related to antifouling improvements.

to their use in order to have at hand important morphological and physical property data. Table 3.2 shows important membrane properties and their effect on performance.

An overview from a variety of sources [31–54] of membrane characterization will be presented in the following sections.

3.5.1 Contact Angle

The concept of contact angle has been introduced with relation to membrane wetting and depicted in Figure 3.2 alongside equation (3.1).

The static contact angle is measured for flat sheet membranes while a dynamic contact angle is measured for hollow fiber membranes. The static contact angle for flat sheet membranes is obtained using the sessile liquid drop method. This has been the most widely used method due to its simplicity and availability of reliable and cost-effective apparatuses. Figure 3.7 shows a contact angle apparatus for flat sheets. The equipment usually comes with an assortment of tools such as syringes for the liquid, membrane holder, camera and a computer with a specialist software. The underlying theory of contact angle measurement is covered in [3].

The dynamic method (also known as the Whilhelmy plate method) can be applied using a commercially available tensiometer. This method is suitable for hollow fiber membranes and the underlying theory is also covered in [3].

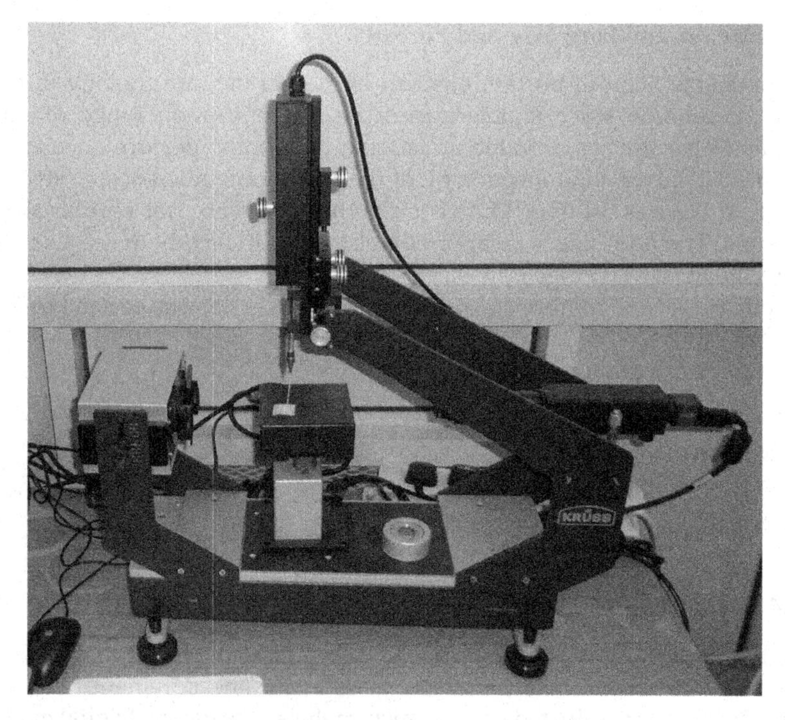

FIGURE 3.7
Contact angle measurement apparatus for flat sheet membranes.

3.5.2 Liquid Entry Pressure

One of the virtues of membrane distillation desalination is operations under mild conditions, namely low temperatures (typically less than 100°C) and low pressure (just enough pressure for liquid flow). If, however, there are reasons to increase slightly the pressure, care must be taken to avoid membrane wetting. This is related to one of the properties of hydrophobic membranes called liquid entry pressure (LEP) or wetting pressure. It is like an upper limit for the liquid pressure and varies from membrane to membrane material. The liquid entry pressure is defined by equation (3.2):

$$LEP = \frac{-2B\gamma_l}{d_{max}}\cos(\theta) \tag{3.2}$$

where B is a geometric factor, γ_l is the surface tension of the liquid solution, θ is the contact angle and d_{max} is the largest pore size of the membrane.

The full description of the experimental description of LEP is covered elsewhere [52,55,56]. Table 3.2 shows the range of values for LEP in the literature.

3.5.3 Membrane Pore Size and Porosity

The pore size is an important element of membrane morphology in membrane distillation since it determines the microfiltration range of operation, and also the actual value affects the membrane performance as seen in Table 3.2. Given the morphology of the membrane microscopic structure shown in Figures 3.4 and 3.5, it is clear that pores are not regular shaped cylinders. The pore size is an approximation and in fact there is a size range with a mean size (average pore size).

There are several methods available for pore size determination [57–59]:

- Gas permeation method
- Wet-dry flow method
- Mercury porosimetry
- Electron microscopy
- Atomic force microscopy

3.5.4 Membrane Thickness

Membrane thickness can be accurately measured using electron microscopy and contactless laser measurements. Reasonable measurements but some errors can also be conducted using a micrometer screw gauge. Digital micrometers can also be used with good accuracy depending on the thickness [60].

3.5.5 Pore Tortuosity

Membranes have irregularly shaped pores and this is described in terms of a property called tortuosity, τ. A straight pore will have a tortuosity of 1. Table 3.2 suggests that values between 1 and 4 have been reported. Tortuosity is estimated indirectly using measurable quantities such as porosity ε and effective pore length ε / L_p where $L_p = \tau\partial$ where ∂ is the membrane thickness [61].

3.5.6 Thermal Conductivity

The thermal conductivity k_m of a MD membrane is a composite value between the polymer material thermal conductivity k_p and the thermal conductivity of gas k_g entrapped in the pores. Both k_p and k_g are readily available in the literature. Table 3.3 shows typical values of thermal conductivities for PTFE, PVDF and PP. The thermal conductivity of gasses is typically reported as 0.02 $Wm^{-1}K^{-1}$.

Depending on the membrane pore structure, the membrane thermal conductivity (sometimes referred to as effective thermal conductivity) can be estimated through calculations using the isostrain or isostress models:

Isostrain model:

$$k_m = \varepsilon\, k_g + (1-\varepsilon)\, k_p \tag{3.3}$$

TABLE 3.3

Thermal Conductivity of Some Hydrophobic Polymers

Polymer	Thermal Conductivity ($Wm^{-1}K^{-1}$)
Polytetrafluoroethylene (PTFE)	0.25
Polyvinylidenefluoride (PVDF)	0.1–0.25
Polypropylene (PP)	0.1–0.22

Isostress model:

$$k_m = \left[\frac{\varepsilon}{k_g} + + \frac{(1-\varepsilon)}{k_p} \right]^{-1} \tag{3.4}$$

Literature reported values for k_m are within the range 0.03–0.06 W m^{-1} K^{-1}.

3.5.7 Concluding Remarks

The base membrane material for membrane distillation has not changed in the past three decades or more and consisted of polytetrafluoroethylene (PTFE) or polyvinyledenefluoride (PVDF) or polypropylene (PP). However, preparation methods evolved significantly toward electrospinning. This technology offers more flexibility to control membrane morphology and associated properties. It also appears that polyvinyledenefluoride membranes are popular in laboratory work due to its solubility at low temperature, making electrospinning easier compared to polymer melt electrospinning of polypropylene.

Commercially available membranes of PVDF, PTFE and PP are also widely used in laboratory work and are provided with measured properties, thus reducing costs in academic experimentation for characterization. Because of the surge in interest in membrane distillation, there appears to be an excessive overlap of results in various journals, as will be seen in the next chapters. Nevertheless, new fields of study with modified PP, PVDF and PTFE membranes are currently ongoing, and this will be covered in a later chapter.

References

1. Smolders, K. and A.C.M. Franken, Terminology for membrane distillation. *Desalination*, 1989. **72**(3): pp. 249–262.
2. Mittal, K.L. et al. *Contact Angle, Wettability and Adhesion*. 2015, Utrecht, the Netherlands: VSP.

3. Yuan, Y. and T.R. Lee, Contact angle and wetting properties, eds. G. Bracco and B. Holst, *Surface Science Techniques: Springer Series in Surface Sciences*. 2013, Berlin, Germany: Springer-Verlag, Vol. **51**, pp. 3–34.
4. Wenzel, R.N., Surface roughness and contact angle. *The Journal of Physical and Colloid Chemistry*, 1949. **53**(9): pp. 1466–1467.
5. Fowkes, F.M., Contact angle, wettability, and adhesion. *Advances in Chemistry*, 1964. **43**; *American Chemical Society* **404**.
6. Zisman, W.A., Relation of the equilibrium contact angle to liquid and solid constitution, in contact angle, wettability, and adhesion. *American Chemical Society*, 1964. **43**: pp. 1–51.
7. Park, J., U. Pasaogullari, and L. Bonville, Wettability measurements of irregular shapes with Wilhelmy plate method. *Applied Surface Science*, 2018. **427**: pp. 273–280.
8. Krishnan, A. et al., An evaluation of methods for contact angle measurement. *Colloids and Surfaces B: Biointerfaces*, 2005. **43**(2): pp. 95–98.
9. Mulder, M., *Basic Principles of Membrane Technology*. 1997, Dordrecht, the Netherlands: Kluwer Academic Publishers.
10. Kitamura, T. et al., Formation mechanism of porous structure in polytetrafluoroethylene (PTFE) porous membrane through mechanical operations. *Polymer Engineering & Science*, 1999. **39**(11): pp. 2256–2263.
11. Lu, J. et al., Effects of diluent on PVDF microporous membranes structure via thermally induced phase inversion. *Technology of Water Treatment*, 2013. **39**: pp. 33–36.
12. Sukitpaneenit, P. and T.-S. Chung, Molecular elucidation of morphology and mechanical properties of PVDF hollow fiber membranes from aspects of phase inversion, crystallization and rheology. *Journal of Membrane Science*, 2009. **340**(1): pp. 192–205.
13. Chandavasu, C. et al., Polypropylene blends with potential as materials for microporous membranes formed by melt processing. *Polymer*, 2002. **43**(3): pp. 781–795.
14. Matsuura, T., *Synthetic Membranes and Membrane Separation Processes*. 1994, Boca Raton, FL: CRC Press.
15. Fath, H.E.S. et al., PV and thermally driven small-scale, stand-alone solar desalination systems with very low maintenance needs. *Desalination*, 2008. **225**(1–3): pp. 58–69.
16. Zakrzewska-Trznadel, G., M. Harasimowicz, and A.G. Chmielewski, Concentration of radioactive components in liquid low-level radioactive waste by membrane distillation. *Journal of Membrane Science*, 1999. **163**(2): pp. 257–264.
17. Ray, S.S. et al., A comprehensive review: Electrospinning technique for fabrication and surface modification of membranes for water treatment application. *RSC Advances*, 2016. **6**(88): pp. 85495–85514.
18. Kim, J.F. et al., Thermally induced phase separation and electrospinning methods for emerging membrane applications: A review. *AICHE Journal*, 2016. **62**(2): pp. 461–490.
19. Brown, T.D., P.D. Dalton, and D.W. Hutmacher, Melt electrospinning today: An opportune time for an emerging polymer process. *Progress in Polymer Science*, 2016. **56**: pp. 116–166.

20. Agarwal, S., *Electrospinning: A Practical Guide to Nanofibers*. 2016, Berlin, Germany: De Gruyter.
21. Mitchell, G.R. (ed.), *Electrospinning: Principles, Practice and Possibilities*. 2015, London, UK: Royal Society of Chemistry.
22. Kao, T.H. et al., Using coaxial electrospinning to fabricate core/shell-structured polyacrylonitrile-polybenzoxazine fibers as nonfouling membranes. *RSC Advances*, 2015. **5**(72): pp. 58760–58771.
23. Ahmed, F.E., B.S. Lalia, and R. Hashaikeh, A review on electrospinning for membrane fabrication: Challenges and applications. *Desalination*, 2015. **356**: pp. 15–30.
24. Jao, C.-S., Y. Wang, and C. Wang, Novel elastic nanofibers of syndiotactic polypropylene obtained from electrospinning. *European Polymer Journal*, 2014. **54**: pp. 181–189.
25. Li, L., R. Hashaikeh, and H.A. Arafat, Development of eco-efficient microporous membranes via electrospinning and annealing of poly (lactic acid). *Journal of Membrane Science*, 2013. **436**: pp. 57–67.
26. Essalhi, M. and M. Khayet, Self-sustained webs of polyvinylidene fluoride electrospun nanofibers at different electrospinning times: 1. Desalination by direct contact membrane distillation. *Journal of Membrane Science*, 2013. **433**: pp. 167–179.
27. Lyons, J., C. Li, and F. Ko, Melt-electrospinning part I: Processing parameters and geometric properties. *Polymer*, 2004. **45**(22): pp. 7597–7603.
28. Bhardwaj, N. and S.C. Kundu, Electrospinning: A fascinating fiber fabrication technique. *Biotechnology Advances*, 2010. **28**(3): pp. 325–347.
29. Simone, S. et al., Effect of selected spinning parameters on PVDF hollow fiber morphology for potential application in desalination by VMD. *Desalination*, 2014. **344**: pp. 28–35.
30. Alsalhy, Q.F. et al., Poly(ether sulfone) (PES) hollow-fiber membranes prepared from various spinning parameters. *Desalination*, 2014. **345**: pp. 21–35.
31. Eykens, L. et al., Characterization and performance evaluation of commercially available hydrophobic membranes for direct contact membrane distillation. *Desalination*, 2016. **392**: pp. 63–73.
32. Xiao, T.H. et al., Fabrication and characterization of novel asymmetric polyvinylidene fluoride (PVDF) membranes by the nonsolvent thermally induced phase separation (NTIPS) method for membrane distillation applications. *Journal of Membrane Science*, 2015. **489**: pp. 160–174.
33. Shirazi, M.M.A. et al., Assessment of atomic force microscopy for characterization of PTFE membranes for membrane distillation (MD) process. *Desalination and Water Treatment*, 2015. **54**(2): pp. 295–304.
34. Qiu, L. et al., Adaptable thermal conductivity characterization of microporous membranes based on freestanding sensor-based 3 omega technique. *International Journal of Thermal Sciences*, 2015. **89**: pp. 185–192.
35. Moradi, R. et al., Preparation and characterization of polyvinylidene fluoride/graphene superhydrophobic fibrous films. *Polymers*, 2015. **7**(8): p. 1444.
36. Bannwarth, S. et al., Characterization of hollow fiber membranes by impedance spectroscopy. *Journal of Membrane Science*, 2015. **473**: pp. 318–326.
37. Zhong, Z. et al., Membrane surface roughness characterization and its influence on ultrafine particle adhesion. *Separation and Purification Technology*, 2012. **90**: pp. 140–146.

38. Prince, J.A. et al., Preparation and characterization of highly hydrophobic poly(vinylidene fluoride)—Clay nanocomposite nanofiber membranes (PVDF–clay NNMs) for desalination using direct contact membrane distillation. *Journal of Membrane Science*, 2012. *397–398*: pp. 80–86.
39. Lv, Y.X. et al., Fabrication and characterization of superhydrophobic polypropylene hollow fiber membranes for carbon dioxide absorption. *Applied Energy*, 2012. **90**(1): pp. 167–174.
40. Hou, D. et al., Preparation and characterization of PVDF/nonwoven fabric flat-sheet composite membranes for desalination through direct contact membrane distillation. *Separation and Purification Technology*, 2012. **101**: pp. 1–10.
41. Essalhi, M. and M. Khayet, Fabrication and characterization of electro-spun nano-fibrous membranes for desalination by membrane distillation. *Procedia Engineering*, 2012. **44**: pp. 235–237.
42. Li, Q.A., Z.L. Xu, and M. Liu, Preparation and characterization of PVDF microporous membrane with highly hydrophobic surface. *Polymers for Advanced Technologies*, 2011. **22**(5): pp. 520–531.
43. Lai, C.L. et al., Preparation and characterization of plasma-modified PTFE membrane and its application in direct contact membrane distillation. *Desalination*, 2011. **267**(2–3): pp. 184–192.
44. Mannella, G.A., V. La Carrubba, and V. Brucato, Characterization of hydrophobic polymeric membranes for membrane distillation process. *International Journal of Material Forming*, 2010. **3**: pp. 563–566.
45. Qtaishat, M. et al., Preparation and characterization of novel hydrophobic/hydrophilic polyetherimide composite membranes for desalination by direct contact membrane distillation. *Journal of Membrane Science*, 2009. **327**(1): pp. 264–273.
46. Hou, D. et al., Fabrication and characterization of hydrophobic PVDF hollow fiber membranes for desalination through direct contact membrane distillation. *Separation and Purification Technology*, 2009. **69**(1): pp. 78–86.
47. Garcia-Payo, M.C., M. Essalhi, and M. Khayet, Preparation and characterization of PVDF-HFP copolymer hollow fiber membranes for membrane distillation. *Desalination*, 2009. **245**(1–3): pp. 469–473.
48. Wyart, Y. et al., Membrane characterization by microscopic methods: Multiscale structure. *Journal of Membrane Science*, 2008. **315**(1–2): pp. 82–92.
49. Grundke, K., Characterization of polymer surfaces by wetting and electrokinetic measurements—Contact angle, interfacial tension, zeta potential, ed. M. Stamm, *Polymer Surfaces and Interfaces: Characterization, Modification and Applications*. 2008, Berlin, Germany: Springer Berlin, pp. 103–138.
50. Wu, B., K. Li, and W.K. Te, Preparation and characterization of poly(vinylidene fluoride) hollow fiber membranes for vacuum membrane distillation. *Journal of Applied Polymer Science*, 2007. **106**(3): pp. 1482–1495.
51. Khayet, M., K.C. Khulbe, and T. Matsuura, Characterization of membranes for membrane distillation by atomic force microscopy and estimation of their water vapor transfer coefficients in vacuum membrane distillation process. *Journal of Membrane Science*, 2004. **238**(1–2): pp. 199–211.
52. Khayet, M. and T. Matsuura, Preparation and characterization of polyvinylidene fluoride membranes for membrane distillation. *Industrial & Engineering Chemistry Research*, 2001. **40**(24): pp. 5710–5718.

53. Khulbe, K.C. and T. Matsuura, Characterization of synthetic membranes by Raman spectroscopy, electron spin resonance, and atomic force microscopy: A review. *Polymer*, 2000. **41**(5): pp. 1917–1935.
54. deZarate, J.M.O., L. Pena, and J.I. Mengual, Characterization of membrane distillation membranes prepared by phase inversion. *Desalination*, 1995. **100**(1–3): pp. 139–148.
55. Franken, A.C.M. et al., Wetting criteria for the applicability of membrane distillation. *Journal of Membrane Science*, 1987. **33**(3): pp. 315–328.
56. Racz, G. et al., Experimental determination of liquid entry pressure (LEP) in vacuum membrane distillation for oily wastewaters. *Membrane Water Treatment*, 2015. **6**(3): pp. 237–249.
57. Khayet, M., J.I. Mengual, and G. Zakrzewska-Trznadel, Direct contact membrane distillation for nuclear desalination. Part I: Review of membranes used in membrane distillation and methods for their characterisation. *International Journal of Nuclear Desalination*, 2005. **1**(4): pp. 435–449.
58. Nakao, S.-I., Determination of pore size and pore size distribution: 3. Filtration membranes. *Journal of Membrane Science*, 1994. **96**(1): pp. 131–165.
59. Cuperus, F.P. and C.A. Smolders, Characterization of UF membranes: Membrane characteristics and characterization techniques. *Advances in Colloid and Interface Science*, 1991. **34**: pp. 135–173.
60. Francis, L. et al., Performance evaluation of the DCMD desalination process under bench scale and large scale module operating conditions. *Journal of Membrane Science*, 2014. **455**: pp. 103–112.
61. Khayet, M., A. Velazquez, and J.I. Mengual, Modelling mass transport through a porous partition: Effect of pore size distribution. *Journal of Non-Equilibrium Thermodynamics*, 2004. **29**(3): pp. 279–299.

4

Membrane Distillation Module Design

4.1 Introduction

In Chapter 2 we saw the various membrane distillation configurations, namely direct contact, air gap, vacuum and sweeping gas configurations. These configurations bear names related to operational methods on the permeate side. Since we also saw in Chapter 3 that membrane distillation membranes can have flat sheet or capillary (hollow fiber) shapes, one can envisage different ways to house these membranes inside the compartment where the distillation process takes place. This compartment is commonly called "module." The membrane distillation module can be rectangular and house a flat sheet membrane in a plate and frame fashion or accommodate a bundle of hollow fibers (capillaries); or it can be tubular and house a bundle of hollow fiber membranes in a shell and tube fashion. Tubular modules can also accommodate a spiral wound [1,2] membrane, but these are rare in membrane distillation compared to reverse osmosis where it is dominant.

We can see a typical flat sheet membrane direct contact membrane distillation module in Figure 4.1. Figure 4.1a depicts a counter-current flow pattern for the feed and permeate, while Figure 4.1b depicts a co-current flow pattern. Figure 4.1c depicts a cross flow pattern (this is not widespread). On the other hand, a hollow fiber DCMD module is shown in Figure 4.2 (a is counter-current flow and b is co-current flow). Direct contact modules are some of the simplest modules to manufacture even though the hollow fiber module needs specialist skills to install the hollow fiber membrane bundle and arrange for inlet/outlet ports. This task is difficult for small diameter hollow fiber membranes.

Figure 4.3 depicts a vacuum membrane distillation module. Clearly, one can see the additional ancillary equipment (vacuum pump, permeate vapor condenser) that are outside the main separation module under vacuum in the permeate side. Despite the higher thermal efficiency of vacuum MD [3], the overall energy efficiency of vacuum MD is often overlooked or ignored because the vacuum side has significant additional energy inputs such as power to the vacuum pump and low temperature condensers often requiring refrigeration to recover as much as possible the permeate. However, a new

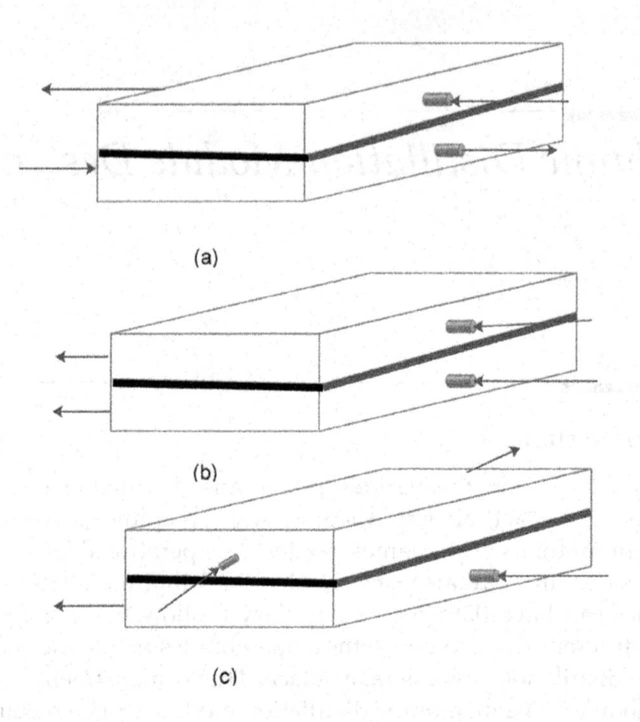

FIGURE 4.1
Typical flat sheet membrane direct contact membrane distillation module – (a) counter-current flow (b) co-current flow (c) cross flow.

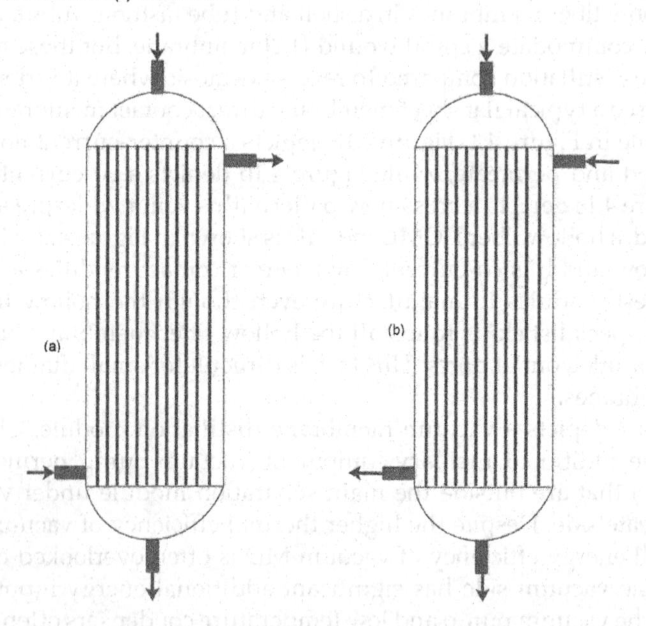

FIGURE 4.2
Hollow fiber DCMD modules – (a) countercurrent flow (b) co-current flow.

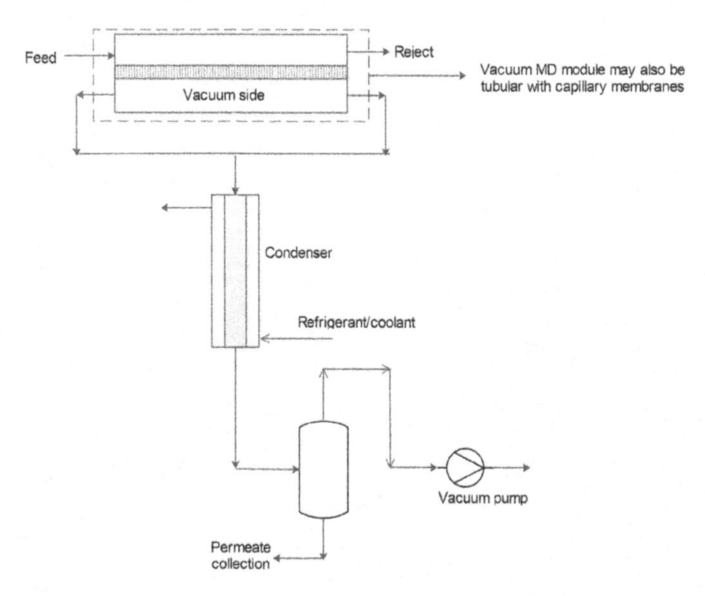

FIGURE 4.3
Vacuum membrane distillation module.

vacuum-multi-effect membrane distillation (V-MEMD) module has been reported [4] where energy efficiency was claimed to be improved by using multiple vacuum effects where thermal energy can be recovered and at the same time using ambient temperature cooling. It has to be said that this new V-MEMD module is commercial (MEMSYS) and the level of sophistication of the module suggests it would not be low cost for the average graduate research projects in MD. However, the MEMSYS system, which is pilot plant scale (roughly 35 L/h on a good sunny day) not laboratory scale, has merit in the sense that it is entirely powered by solar energy (12 solar thermal collectors with average thermal energy production of 80 kWh/day and 1.5 kW peak electricity by 6 solar cells), and this definitely enhances the energy efficiency, not just the thermal efficiency [4]. What was said here is of particular significance when it comes to comparing laboratory based thermal/energy efficiencies with pilot plant or large-scale plant efficiencies.

In Figure 4.4, we can see the sweep gas membrane distillation module and its ancillary equipment (gas compressor, typically air for desalination, a permeate vapor condenser and storage tank). Since this configuration was not investigated to the same extent as other MD configurations, it is important to give some weight to recent reporting on SGMD module performance. Zhao et al. found that the sweeping gas flowrate has a significant impact on the permeate vapor mass transfer rate [5]. This obviously would have implications on the energy input on the permeate side. They also found that at low sweeping gas flowrates, the gas could be saturated with vapor with potential or even probable deposition of permeate droplets on the membrane on the

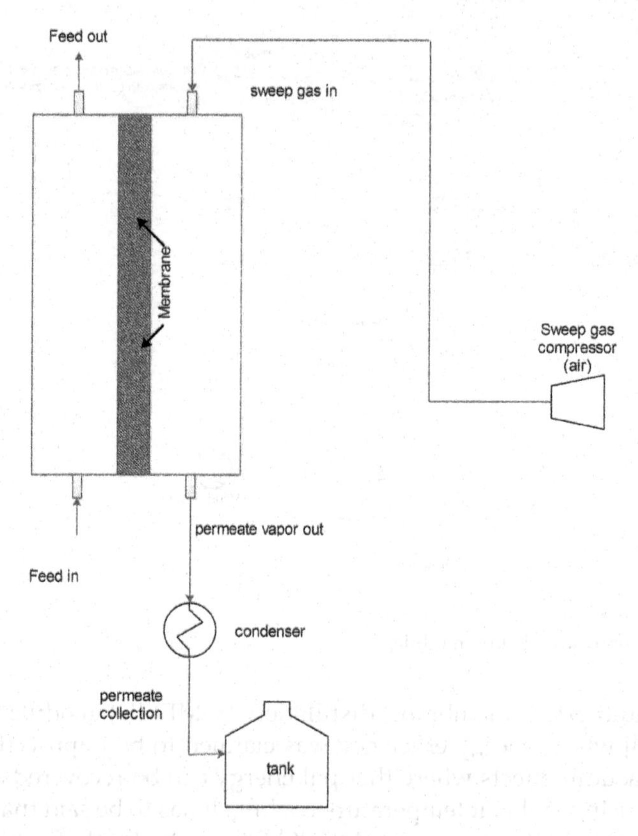

FIGURE 4.4
Sweep gas membrane distillation module.

permeate side. These droplets would evaporate eventually and be swept by the sweeping gas. This phenomenon has not been quantified in the literature in terms of modeling and may well have an impact on flux performance of SGMD modules.

Finally, Figure 4.5 depicts an air gap membrane distillation module. The permeate side incorporates a cold plate adjacent to an air gap to enable permeate liquid collection. This configuration has been widely used in MD but not a great deal was published on the effects of the extent of the air gap, especially when the membrane deformation (due to flow pressure however small it may be on the feed) can impact the gap cavity. Some researchers stabilized the membrane with a wire mesh [6]. In addition, the permeate collection on a cooled plate requires low temperatures that may require some refrigeration, adding to the energy input in the permeate side. In a paper, Khayet and Cojocaru [7] described a specific performance index of an AGMD that can be estimated using all forms of electrical power input into the AGMD system, corresponding to a specific energy consumption of 5.3 kWh/m^3. This specific energy consumption contrasts the greatly higher

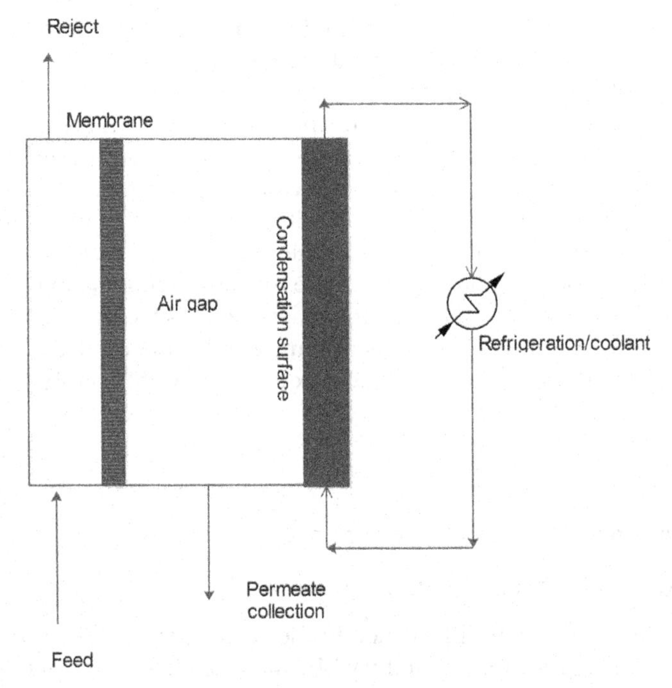

FIGURE 4.5
Air gap membrane distillation module.

value obtained of 810 kWh/m^3 obtained by Guillén-Burrieza et al. [8] on an AGMD pilot plant. Again, this wide difference arises possibly on the scale effect between laboratory and pilot plant systems.

Regardless of the membrane distillation configurations described in Chapter 2, certain operational parameters must be strictly observed in order to ensure a meaningful justification for the application of membrane distillation for desalination. These important operational parameters include feed temperatures that should not exceed 80°C (if low-grade waste heat [9–11] and solar stills energy [12–17] were to be used) and liquid feed pressure and pressure drop so that the membrane used does not experience a pressure greater than its specific liquid entry pressure (LEP) limit to avoid pore wetting [18–34]. In order to control and record temperatures, a minimum level of instrumentation is required. The most important instruments (not shown in Figures 4.1 through 4.5) include temperature and pressure sensors for the feed line and at least temperature sensors for the permeate line. In addition, means to measure and control flowrates are also highly desirable for serious research or development work.

The design of high-performance membrane distillation modules within these limits can prove to be challenging for commercial deployment perspectives. Indeed, one has to acknowledge that the bulk of research output on membrane distillation was conducted on small-scale laboratory setups with operational parameters that cannot be easily scaled up and translated into real units thermally integrated with low-grade waste heat sources [35,36] or outdoor solar collectors [37,38].

In this chapter the basic geometric and flow pattern impacts on module designs will be explored alongside methods to enhance module designs so that membrane distillation performance can be enhanced and made within reach to larger scale deployment where it matters most, namely integrated with low-grade waste heat sources or remote solar stills. Let us also remind ourselves that design in the broader sense in engineering means "sizing and giving shapes" to functional units to achieve some specific output in quantity and quality. In the context of membrane distillation for desalination, the design target would be the permeate flux and permeate quality (conductivity, TDS).

4.2 Module Geometric Considerations

Let us illustrate the concept of membrane distillation module design and fabrication through simple laboratory MD modules. The effect of the module design on flux performance will be detailed in the next sections.

A flat sheet membrane direct contact module consists of two separate channel compartments (One for the hot feed and the other for the cold permeate) machined from blocks of solid polymer (PTFE, PP, etc… as long as the polymer has good chemical resistance and low thermal conductivity). The module design gives flexibility on the flow channel dimensions (length, width and depth) depending on the membrane size and range of flows anticipated to cover a wide enough range, especially for research purposes. One of the challenges would be to cover turbulent flows with reasonable fluid velocities. One way to accomplish this is to use turbulence promoters, also called "spacers." Spacers are widely used in compact modules to ensure an adequate degree of turbulence is achieved. Unfortunately, it is currently only possible to "assume" that turbulence takes place unless sophisticated flow visualization is undertaken using exact replicas of the module but made of transparent material, and using colored tracers, one can visualize the flow. Other means of studying flow using the actual modules is through the use of ionic tracers to measure the conductivity downstream (outlet). These methods have not been documented in the literature even though they can actually enrich the experimental and associated modeling work.

Let us see typical laboratory DCMD module compartments showing the feed and permeate channels and the associated membrane. Commercial membranes can be purchased as readymade coupons of particular dimensions or as rolls from which coupons can be cut to the desired dimensions to fit the module compartments. Figure 4.6 depicts a set of compartments

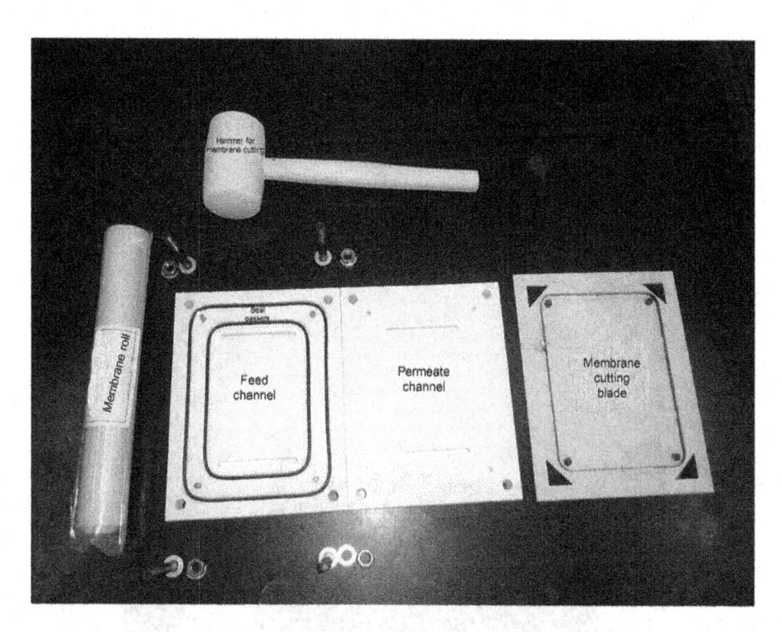

FIGURE 4.6
Feed and permeate flow channels in a DCMD module with homemade membrane cutting blade.

representing flow channels (feed and permeate) fitted with "gasket" seals and a tool for cutting the membrane from commercial roll. Figure 4.7 depicts the cut membrane extraction. The membrane is then "sandwiched" between the compartments, a spacer can be inserted on the top of the membrane in the feed side (or both sides) and the whole assembly secured with nuts and bolts. A spacer can be seen in Figure 4.8.

The assembled module is shown in Figure 4.9. We can actually see one of the ports in which the tubing for feed or distillate can be attached to the corresponding tank.

Now, let us turn our attention to a hollow fiber module. There are two components in the hollow fiber module: the hollow fiber bundle depicted in Figure 4.10 and the shell that houses the hollow fiber bundle, depicted in Figure 4.11. Actually, Figure 4.11 shows two hollow fiber modules, one of them covered with an insulating foam to minimize heat losses due to the thin nature of the shell.

In a membrane distillation laboratory, the module is only one unit out of several that constitute the full set-up. Even though the module is the heart of the membrane distillation process, often it can be quite small in appearance compared to the rest of the ancillary equipment and instrumentation. Let us see an example of a laboratory DCMD setup (hollow fiber direct contact module). This is shown in Figure 4.12. The setup consists of the insulated hollow fiber membrane module and on either side, we can see the feed tank or permeate collection tanks (resting on a digital balance), the feed pump or permeate pump, the feed heater or the permeate cooler, and an assortment

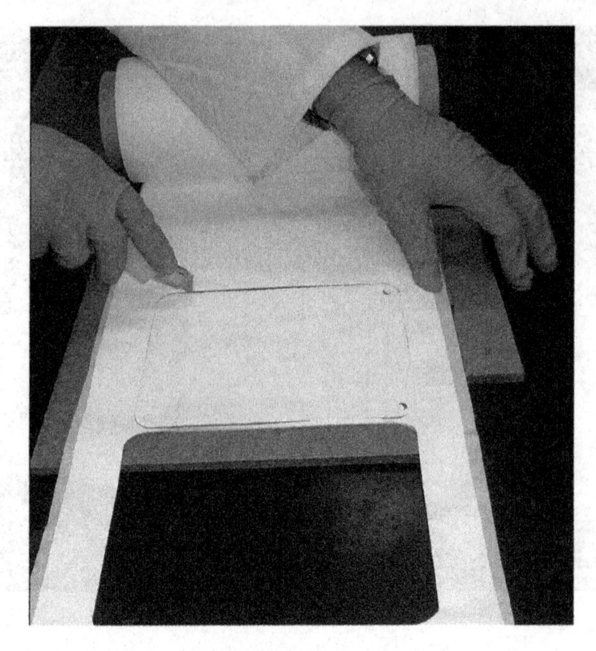

FIGURE 4.7
Flat sheet membrane extraction after cutting to measure for fitting in a membrane distillation module.

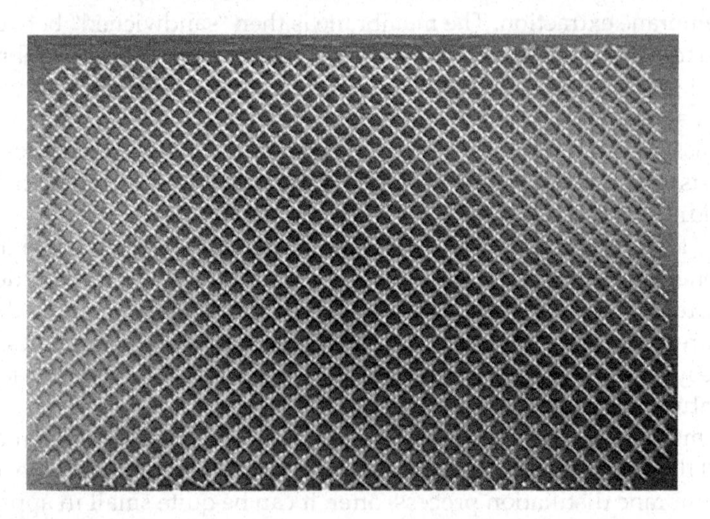

FIGURE 4.8
Spacer for a purpose made flat sheet membrane distillation module.

FIGURE 4.9
DCMD module assembly showing one of the connection ports.

FIGURE 4.10
Hollow fiber membrane bundle for membrane distillation.

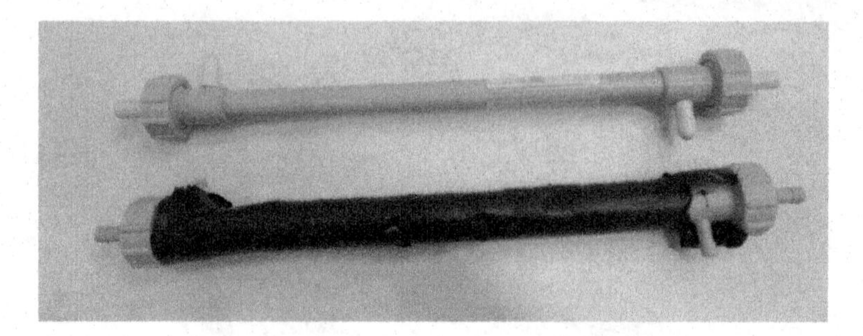

FIGURE 4.11
Shells (bare and insulated) of direct contact hollow fiber membrane modules.

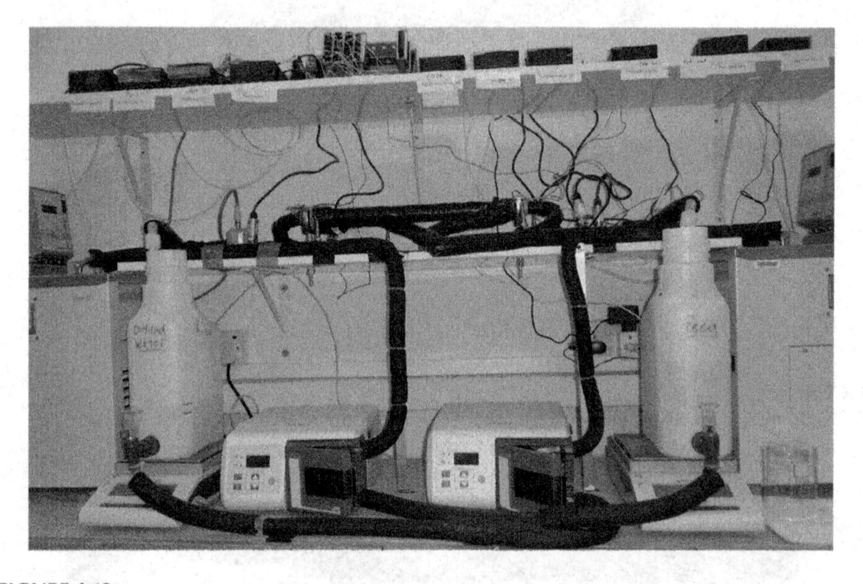

FIGURE 4.12
Laboratory set-up for direct contact membrane distillation in desalination, including hollow fiber module, ancillary equipment and instrumentation/control.

of sensors (flowmeter, temperature, pressure) connected to digital displays on the top shelf. In addition to visual display of process variables (flowrates, temperatures and pressures), one can also add simple data acquisition to log all variables in real time to a computer for further analysis of data. For research work, the data acquisition is important for both steady state or dynamic experiments.

The above discussion on the design and fabrication of a MD module is quite useful for beginners.

Now, let us see how the module design can affect the permeate flux.

4.2.1 Rectangular Modules

The effects of membrane and module design on permeate flux, through theoretical and experimental considerations, were demonstrated by Martinez and Rodriguez-Maroto [39]. They started their argument on well defined and known membrane distillation flux and associated equations. The flux is given by equation (4.1):

$$J = C\left(P_{m1} - P_{m2}\right) \tag{4.1}$$

where J is the distillate (permeate) flux, C is the membrane coefficient (almost dependent on the membrane properties only), P_{m1} and P_{m2} are the vapor pressures on the feed and permeate side respectively. Assuming the vapor transport through the membrane porous matrix is approximated by the Dusty Gas Model [40–44] with a combined Knudsen-molecular mechanism, the membrane coefficient C can be defined as [39]:

$$C = \frac{\varepsilon_s}{q\delta}\frac{M}{RT}\left(\frac{1}{D_K} + \frac{P_a}{pD_{wa}}\right)^{-1}_{\log mean} \tag{4.2}$$

where ε_s is the superficial porosity of the membrane, δ the membrane thickness, q the tortuosity factor of the membrane pores, M the molecular weight of water, R the gas constant, T the temperature, P_a the partial pressure of the air entrapped in the pores, p the total pressure inside the pores, D_K is the Knudsen diffusion coefficient of water vapor and D_{wa} is the diffusion coefficient of water in air.

The partial pressures P_{m1} and P_{m2} need to be estimated at the membrane surface temperatures T_{m1} and T_{m2}. These membrane surface temperatures can be estimated from equations (4.3 and 4.4) [39]:

$$T_{m1} = T_{b1} - \left(T_{b1} - T_{b2}\right)\frac{\frac{1}{h_1}}{\left(\frac{1}{H}\right) + \left(\frac{1}{h}\right)} \tag{4.3}$$

$$T_{m2} = T_{b2} - \left(T_{b1} - T_{b2}\right)\frac{\frac{1}{h_2}}{\left(\frac{1}{H}\right) + \left(\frac{1}{h}\right)} \tag{4.4}$$

where T_{b1} and T_{b2} are the bulk temperatures in the feed and permeate channels, and

$$H = \frac{k_m}{\delta} + \frac{J\Delta_v}{T_{m1} - T_{m2}} \tag{4.5}$$

H being the effective thermal conductivity of the membrane and

$$\frac{1}{h} = \frac{1}{h_1} + \frac{1}{h_2} \tag{4.6}$$

where h is the global film mass transfer coefficient with h_1 and h_2 defined in a parametric form [39]:

$$h_i = Ak_i\left(\frac{\rho_i v}{\mu_i}\right)^{\alpha}\left(\frac{c_p\ \mu_i}{k_i}\right)^{0.33} \quad i \text{ is } 1 \text{ or } 2 \tag{4.7}$$

and the feed concentration on the membrane surface is given by [39]:

$$x_{m1} = x_{b1}\exp\left(\frac{J}{\rho K}\right) \tag{4.8}$$

with

$$K = AD_1\left(\frac{\rho_1 v}{\mu_1}\right)^{\alpha}\left(\frac{\mu_1}{\rho_1 D_1}\right)^{0.33} \tag{4.9}$$

where k is the fluid thermal conductivity (subscript 1 is feed side, and 2 for permeate side); D is the solute diffusion coefficient; μ is the fluid viscosity; c_p the fluid specific heat; v the fluid mean linear velocity on membrane surface; A is a parameter including geometric characteristics of the membrane module and the value of α is to be determined by the state of development of the velocity, temperature and concentration profiles along the flow channel in the membrane distillation module [39].

Martinez and Rodriguez-Maroto [39] fitted their experimental data to equations (4.2 through 4.9) and obtained the following results for the unknown parameters in equations (4.7 through 4.9):

α (flow channel without spacer) = 0.45, α (channel with spacer) = 0.72

A (flow channel without spacer) = 8.22, A (channel with spacer) = 0.90

In this elegant demonstration of the effect of operational parameters on module design, it is important to bear in mind that the module dimensions (such as flow channel characteristic dimensions) are implied in the calculations of the liquid velocities and the membrane active area (for the flux calculation).

4.2.2 Cylindrical Modules

The cylindrical module normally houses a bundle of hollow fiber membranes, traditionally installed in a shell and tube fashion with straight tubes [45,46]. However, the literature on hollow fiber membrane modules described hollow fiber tube arrangements other than straight tubes [47–49]. On rare occasions, spiral membranes housed in cylindrical modules were reported [50–52].

For desalination or brine concentration applications, the saline feed would normally be located in the shell side and permeate collected in the tube side because of the high potential of membrane fouling/scaling. However, in some MD configurations such as vacuum MD, the vapor permeate was reported to be removed from the shell side by the vacuum drawn by an external pump [53]. The allocation side (shell or tube side) of vapor permeate in desalination using hollow fiber membranes in cylindrical modules is often overlooked, especially when experiments last a long period where membrane scaling and wetting is likely to occur. The potential for membrane wetting in vacuum membrane distillation is reported to be "high" [54], and yet the feed was reported to be in the capillary side, presumably because of the potential hollow fiber collapse under vacuum, although this was not explicitly stated [55]. It is therefore important to check if the experiments conducted on hollow fiber membrane modules involve saline feed or pure water and then reconciling this fact with the allocation of saline feed to either the shell or tube side. At this point it is important to bear in mind that the design of cylindrical modules with imbedded bundles of hollow fiber membranes is not trivial nor easy and requires considerable experience to ensure the system is leak free and that the bonding of the membrane bundle to the cylindrical shell will not be affected by membrane "wicking" of the sealant/glue that could lead to pore blockages which will be difficult to detect once the module is sealed. Most of the literature on cylindrical modules with hollow fiber membranes does not cover detailed design and assembly procedure of the capillaries bundle. Often, the assembly is permanently sealed to prevent leakage and by default, this means the membranes cannot be checked, maintained or inspected until the module is opened by destroying the permanent seal, thus rendering further reuse of the module impossible.

The cylindrical shape of the module was also reported for air gap MD. The work was, however, done using only one single hollow fiber membrane positioned in the center of the concentric annulus for the cold water flow [56].

The literature consistently reported that permeate fluxes in conventional straight capillary tubes module are lower than fluxes in flat sheet membrane modules [46,57,58]. Consequently, there have been efforts to investigate why flat sheet membrane modules were superior to straight tube hollow fiber membrane modules and how permeate flux can be enhanced in the latter MD configuration. A number of factors were attributed to the lower flux in hollow fiber modules [48] and these included flow maldistribution [59] and lower convective heat transfer coefficients [46]. We will explore an extract in the literature on studies aimed at improving hollow fiber membrane module performance.

Yang et al. [48] reported on work about five types of hollow fiber module configurations where the capillary tubes were structured-straight, curly, central-tubing for feeding, spacer-wrapped and spacer-knitted. They used direct contact membrane distillation to conduct flux experiments, fluid dynamics studies, and tracer-response tests for flow distribution as well as process heat transfer analysis. Their results showed flux enhancement from 53% to 92% compared to the conventional module design. The spacer-knitted module was reported to have the best performance. Yang et al. [48] also reported that the fluxes in all modified configurations, with the exception of the structured-straight module, were independent of the feed flow velocity, and that the modules with undulating membrane surfaces, namely, curly and spacer-knitted fibers, achieved more than 300% flux improvement in the laminar flow regime. They attributed the improved performance to improved fiber geometries or arrangements that can provide effective boundary layer surface renewal and more uniform flow distribution. That was confirmed by their sodium chloride tracer response measurements. With respect to heat transfer analysis, they found that the advantage of the module with curly fibers led to the least temperature polarization effect and that obviously was favorable for flux enhancements.

Ali et al. [60] studied the effect of different hollow fiber membrane configurations and flow patterns on performance in membrane distillation. The modules they used had helical and wavy conformations. These tested under various hydrodynamic conditions and their performance was compared to conventional straight fiber modules. The effect of flow patterns was also investigated by applying intermittent and pulsating flows to straight hollow fiber membranes. They observed flux enhancements of 47% and 52% relative to straight fibers with the helical and wavy configurations respectively even though packing density of such modules was deemed significantly less than their straight counterparts. For intermittent flow, an improvement of about 30% was observed. The difference was said to be more prominent at low flow rates and approached that of the straight fiber performance under steady flow at high Reynolds numbers for all hollow fiber configurations and flow patterns considered in the research work. The intermittent flow and wavy fibers indicated an energy efficiency improvement of around 180% and 90% over their conventional counterparts respectively. The surface, volume-based enhancement factor and packing density for the intermittent flow and pulsating flow cases were deemed to be optimum.

García-Fernández et al. [61] investigated the effect of the corrugation size and shape of hollow fiber (membranes prepared using wet/wet spinning technique and a micro-engineered spinneret or spraying the external coagulant by micro-jet nozzles) on the performance of direct contact membrane distillation using pure water and sodium chloride solution as feed. The corrugated outer surface acted as micro-turbulence promoters mitigating the temperature polarization effect and enhanced the external effective surface area for condensation. Both factors improved the DCMD permeability of the hollow fiber membranes. However, they found that corrugations with V-shaped valley depths greater than about 30 μm did not always improve the permeate flux. In addition, the membrane prepared with the spray wetting mode exhibited the best desalination performance.

Aryapratama et al. [62] developed a hollow fiber air gap membrane distillation module equipped with multiple cooling channels made of stainless steel. The performance of the AGMD module was evaluated by conducting experiments where the effect of feed temperature and flow rate, ratio of membrane surface area to condensation surface area, membrane packing position were compared with hollow fiber direct contact membrane distillation. Their work showed that the flux and thermal efficiency were up to 12.5 kg/(m^2h) and 81.7% respectively.

Mejia Mendez et al. [63] proposed the use of a thin, self-standing dense membrane instead of microporous materials as a solution to avoid wetting in membrane distillation for water desalination. They first developed a membrane contactor model based on mass and energy balances and validated simulations using experimental data from the literature on hollow fiber modules with microporous membranes. They then performed a comparative analysis of results between a porous and a dense membrane hollow fiber module. A parametric study on the influence of dense membrane thickness and water permeability showed that a two-fold increase in water flux, without significant impact on energy efficiency, was potentially achievable with thin and permeable dense materials compared to microporous membranes. They also provided guidelines for the design of high-performance dense membrane modules for membrane distillation. This study is a rare work on the potential of dense membranes for membrane distillation and more work is needed in the laboratory to ascertain not only the suitability of such membranes for MD desalination, but also their stability and wetting resistance, which was the basis of the claims of Mejia Mendez et al. [63].

Singh et al. [64] claimed to have developed successfully a hollow-fiber membrane (HFM) based rectangular module with hot brine in cross-flow over the HFMs and obtained high and stable water flux under demanding scaling conditions. Their module design with a low surface area/device volume was said to be inadequate for larger-scale plants. A novel cylindrical cross-flow module containing high-flux composite hydrophobic HFMs was described for membrane surface areas, 0.15 and 0.6 m^2, using porous fluorosiloxane-coated porous polypropylene hollow fibers. The simple and easily-scalable module design they developed was claimed to pack four times membrane surface area/unit equipment volume compared to earlier designs. Distilled water production rates from 1 wt% saline feed were studied over a range of temperatures of 60°C–91°C.

4.3 Novel Module Configurations

New MD configurations including multi-stage and multi-effect membrane distillation (MEMD), vacuum multi-effect membrane distillation (VMEMD), hollow fiber multi-effect membrane distillation, material gap membrane distillation (MGMD) as well as unconventional membrane geometries were described in a review by Wang and Chung [65].

Novel module configuration (other types not commonly used like permeate gap [PGMD]) were reported [66] and material gap MDMD [67].

Alsaati and Marconnet [68] argued that localizing heat generation at the liquid-membrane interface in membrane distillation could reduce energy consumption while maintaining sufficient mass flux. In their paper, a locally heated membrane distillation approach was analytically and experimentally evaluated. The measured evaporated mass flux matched the analytical limits considering diffusive and advective mass transfer across a heated porous membrane into stagnant air. Experimentally, thermally-stable silver membranes demonstrate similar mass flux compared to conventional polymer-based membrane at the same surface temperature. At high membrane temperatures, namely at 80°C, the locally heated membrane distillation demonstrates good efficiency with up to 75% reduction of energy compared to direct contact membrane distillation and air gap membrane distillation. They also argued that efficiency can be further improved with better thermal design of the supporting structure and permeate heat recovery. This approach is obviously at an early stage of development and more work needs to be done, especially for desalination applications.

Huang et al. [69] described in a recent paper how to efficiently collect vaporized and condensed water using immobilization of graphene-based material on hydrophobic PTFE membrane surface for water desalination via photothermal membrane distillation. The authors claimed that the water vapor flux of pDA-rGO/PTFE membrane was as much as 78.6% higher than for the bare PTFE membrane case. This work is obviously part of the so-called innovative membranes for future more effective membrane distillation applications, and as such more work is needed to push the limits of confidence further in such improved membranes.

4.4 Fluid Dynamics and Heat Transfer Considerations: Qualitative Considerations

Since membrane distillation modules are flow channels where complex flow patterns can arise as a result of the internal geometry or with the incorporation of turbulence promoting inserts such as spacers, there are opportunities to gain an insight into these complex flows through computational fluid dynamics (CFD). Indeed, CFD is a remarkable simulation tool that is advancing fast and being used in an increasing number of processes to optimize mechanical designs that affect flow patterns which in turn affect transport phenomena. The main advantages of CFD include the possibility to visualize, numerically speaking, the flow and pressure fields in complex geometries without resorting to costly or virtually impossible actual experimental visualizations. In many cases it is simply impossible to observe flows when the material of the flow enclosure is not transparent, hence eliminating optical access to phenomena inside. Fortunately, CFD tools have evolved in the past few decades to offer new opportunities to explore flow fields in membrane distillation channels. It seems the interest in applying CFD to MD module design is also growing [70–84]. Much of the CFD work in MD so far consists of interpreting observed results, improving the spacer geometry to enhance turbulence (and indirectly thermal performance) and optimizing mechanical aspects of module design. Being a relatively recent tool applied to MD, a great deal of work needs to be done. In particular, there is so far no "built-in" physics for membrane distillation in any commercial or open source CFD tools. This has caused some discomfort to the research community who could not find sufficient details on the physics of MD in CFD simulations of MD operations. This drawback remains a point of weakness on using CFD for MD. Another drawback that applies to almost all applications is the computing power required to study very complex MD geometries without resorting to oversimplification or making unrealistic assumptions. Unfortunately, these issues do not help much the case for more CFD applications. Nevertheless, the growing number of publications on CFD applications to MD can be considered as a sign that the situation may improve in the future. Let us explore an extract of the literature where CFD was used in MD.

Cipollina et al. [71] indicated in their paper that only limited work dealt with the problem of MD modules optimization. Their work was aimed at the CFD simulation of the fluid flow and temperature fields within spacer-filled MD module channels for a variety of spacer geometries. They indicated that commercial and custom-made spacer geometries were simulated in order to identify the most important parameters affecting process efficiency.

Soukane et al. [85] published a recent paper on CFD for MD and indicated that the aim of the numerical investigation was to suggest the best modeling strategy that will enable innovation; in other words, modify or check new direct contact MD module designs or operation in a timely manner. Their paper also shed light on the CPU time used to solve cases where convective or conjugate heat transfer should be adopted in the CFD models as well as the CPU time penalty incurred for 3D simulations and any benefits that may ensue in investing in long computational times. More details can be found in their excellent paper [85].

Ghaleni et al. [86] used CFD to investigate the significance of geometrical and physical parameters of hollow fiber membrane modules in the membrane distillation process because they claimed that it has not been fully evaluated. They developed a three-dimensional, multi-physics model of a hollow fiber membrane module aimed at investigating the effect of operating and design parameters on the module performance. The permeate flux and thermal efficiency of the system were considered as the characteristic parameters of the module operated in direct contact mode. The paper offered a good coverage of geometry meshing and how to simplify the Dusty Gas Model to simulate the vapor transport across the membrane. Ghaleni et al. [86] adopted classic MD equations and correlations in their CFD model and used the commercial software COMSOL Multiphysics 5.2a installed on one computing node that had 16 CPU cores (3.2 GHz) and 128 GB RAM, which is considered powerful enough for 3-D simulations.

4.5 Practical Considerations

New and inexperienced researchers tend to be inspired by published work descriptions of MD modules, and without doing essential theoretical design sizing through computations first, results can indeed be disastrous and costly. A good start to module design for research would typically be to consider whether the module will be for flat sheets or hollow fiber membranes or both and for what aims. If the work involves membrane fouling/scaling, what would be the more reliable to extract used membranes without causing damage for inspection and autopsy? One of the most common problems in MD modules is leaks. If the module is poorly designed and allows leaks, however small they may be, there will be a risk of losing valuable results and repeating runs after fixing the leaks. This is the most dreaded problem of graduate students who have limited time for experimental work.

Another potential problem often overlooked by inexperienced researchers is the range of flow regimes possible once the module channels are made. So, module sizing is absolutely crucial and should be among the first determinations before machining or cutting the module casing.

For flat sheet membranes, the machining and grooves for channels and gaskets should be done by professional technicians in order to have the dimensions required and to have a tight gasket fit.

Hollow fiber modules are the most difficult to assemble, especially for high packing density bundles. Often, the capillary tubes bundle needs to be glued to the empty shell and most of the time problems arise in this delicate operation. Choice of resins is also critical to ensure a robust assembly and also to avoid membrane "wicking" of glue/solvent that will cause unexpected operational problems later. That is probably the reason why more researchers opt for commercially available hollow fiber cartridges even if it means being constrained by the sizes available and the packing density of the bundle.

Finally, the MD module is often the smallest item in the MD setup. As can be seen in Figure 4.12, the space footprint of the MD module is a tiny fraction of the entire MD setup where ancillary equipment and instrumentation takes the most laboratory space. Hence it is important to plan carefully the entire setup rather than just the module.

4.6 Concluding Remarks

The design of MD modules is very important and can determine the performance to a great extent. With so much material being published in the literature, it should nowadays be easier to avoid costly errors in fabricating MD modules. With the advent of CFD tools, it should be possible to explore a number of options with respect to module mechanical design to achieve an optimum design that would provide the services expected within the operational parameters anticipated. Given that membrane fouling and scaling still continues to be an issue before commercial deployment is a reality on a large scale, it is important to design modules that can facilitate membrane inspection and recovery without causing damage to the used membrane if it is to be autopsied. The design of hollow fiber membrane modules still needs ways to open the casing and extract the hollow fiber bundle without breaking irreversibly the ends of the module. Overcoming this design issue will make hollow fiber membrane modules at the forefront of research for large scale deployment because of the compactness of these modules.

The new module configurations described above (multi-stage and multi-effect membrane distillation [MEMD], vacuum multi-effect membrane distillation [VMEMD], hollow fiber multi-effect membrane distillation, material gap membrane distillation [MGMD]) still need more work in terms of optimum design to reduce costs if they prove to have economically superior value.

References

1. Ruiz-Aguirre, A. et al., Modeling and optimization of a commercial permeate gap spiral wound membrane distillation module for seawater desalination. *Desalination*, 2017. **419**: pp. 160–168.
2. Chang, H.A. et al., Modeling and optimization of a solar driven membrane distillation desalination system. *Renewable Energy*, 2010. **35**(12): pp. 2714–2722.
3. Zhang, Y. et al., Review of thermal efficiency and heat recycling in membrane distillation processes. *Desalination*, 2015. **367**: pp. 223–239.

4. Zhao, K. et al., Experimental study of the memsys vacuum-multi-effect-membrane-distillation (V-MEMD) module. *Desalination*, 2013. **323**: pp. 150–160.
5. Zhao, S.F. et al., Condensation studies in membrane evaporation and sweeping gas membrane distillation. *Journal of Membrane Science*, 2014. **462**: pp. 9–16.
6. Dehesa-Carrasco, U., C.A. Perez-Rabago, and C.A. Arancibia-Bulnes, Experimental evaluation and modeling of internal temperatures in an air gap membrane distillation unit. *Desalination*, 2013. **326**: pp. 47–54.
7. Khayet, M. and C. Cojocaru, Air gap membrane distillation: Desalination, modeling and optimization. *Desalination*, 2012. **287**: pp. 138–145.
8. Guillén-Burrieza, E. et al., Experimental analysis of an air gap membrane distillation solar desalination pilot system. *Journal of Membrane Science*, 2011. **379**(1–2): pp. 386–396.
9. Dow, N. et al., Pilot trial of membrane distillation driven by low grade waste heat: Membrane fouling and energy assessment. *Desalination*, 2016. **391**: pp. 30–42.
10. Ammar, Y. et al., Desalination using low grade heat in the process industry: Challenges and perspectives. *Applied Thermal Engineering*, 2012. **48**: pp. 446–457.
11. Luo, A. et al., Mapping potentials of low-grade industrial waste heat in Northern China. *Resources, Conservation and Recycling*, 2017. **125**: pp. 335–348.
12. Arunkumar, T. et al., A review of efficient high productivity solar stills. *Renewable and Sustainable Energy Reviews*, 2019. **101**: pp. 197–220.
13. Kaviti, A.K., A. Yadav, and A. Shukla, Inclined solar still designs: A review. *Renewable and Sustainable Energy Reviews*, 2016. **54**: pp. 429–451.
14. Panchal, H. and I. Mohan, Various methods applied to solar still for enhancement of distillate output. *Desalination*, 2017. **415**: pp. 76–89.
15. Zhou, J. et al., Performance analysis of solar vacuum membrane distillation regeneration. *Applied Thermal Engineering*, 2018. **144**: pp. 571–582.
16. Abdallah, S.B., N. Frikha, and S. Gabsi, Simulation of solar vacuum membrane distillation unit. *Desalination*, 2013. **324**: pp. 87–92.
17. Koschikowski, J., M. Wieghaus, and M. Rommel, Solar thermal-driven desalination plants based on membrane distillation. *Desalination*, 2003. **156**(1): pp. 295–304.
18. Li, L. and K.K. Sirkar, Studies in vacuum membrane distillation with flat membranes. *Journal of Membrane Science*, 2017. **523**: pp. 225–234.
19. Hagedorn, A. et al., Membrane and spacer evaluation with respect to future module design in membrane distillation. *Desalination*, 2017. **413**: pp. 154–167.
20. Eykens, L. et al., How to optimize the membrane properties for membrane distillation: A review. *Industrial & Engineering Chemistry Research*, 2016. **55**(35): pp. 9333–9343.
21. Zmievskii, Y.G., Determination of critical pressure in membrane distillation process. *Petroleum Chemistry*, 2015. **55**(4): pp. 308–314.
22. Shim, W.G. et al., Solar energy assisted direct contact membrane distillation (DCMD) process for seawater desalination. *Separation and Purification Technology*, 2015. **143**: pp. 94–104.
23. Chen, Z.L. et al., Study on structure and vacuum membrane distillation performance of PVDF membranes: II. Influence of molecular weight. *Chemical Engineering Journal*, 2015. **276**: pp. 174–184.
24. Yang, Y.F. et al., Criteria for the selection of a support material to fabricate coated membranes for a life support device. *RSC Advances*, 2014. **4**(73): pp. 38711–38717.

25. Racz, G. et al., Theoretical and experimental approaches of liquid entry pressure determination in membrane distillation processes. *Periodica Polytechnica-Chemical Engineering*, 2014. **58**(2): pp. 81–91.
26. Hwang, H.J. et al., Direct contact membrane distillation (DCMD): Experimental study on the commercial PTFE membrane and modeling. *Journal of Membrane Science*, 2011. **371**(1–2): pp. 90–98.
27. He, K., H.J. Hwang, and I.S. Moon, Air gap membrane distillation on the different types of membrane. *Korean Journal of Chemical Engineering*, 2011. **28**(3): pp. 770–777.
28. Khayet, M. et al., Design of novel direct contact membrane distillation membranes. *Desalination*, 2006. **192**(1–3): pp. 105–111.
29. Khayet, M., J.I. Mengual, and G. Zakrzewska-Trznadel, Direct contact membrane distillation for nuclear desalination. Part I: Review of membranes used in membrane distillation and methods for their characterisation. *International Journal of Nuclear Desalination*, 2005. **1**(4): pp. 435–449.
30. Khayet, M. et al., Preparation and characterization of polyvinylidene fluoride hollow fiber membranes for ultrafiltration. *Polymer*, 2002. **43**(14): pp. 3879–3890.
31. Garcia-Payo, M.C., M.A. Izquierdo-Gil, and C. Fernandez-Pineda, Wetting study of hydrophobic membranes via liquid entry pressure measurements with aqueous alcohol solutions. *Journal of Colloid and Interface Science*, 2000. **230**(2): pp. 420–431.
32. Kim, B.-S. and P. Harriott, Critical entry pressure for liquids in hydrophobic membranes. *Journal of Colloid and Interface Science*, 1987. **115**(1): pp. 1–8.
33. Khayet, M., Membranes and theoretical modeling of membrane distillation: A review. *Advances in Colloid and Interface Science*, 2011. **164**(1–2): pp. 56–88.
34. Alkhudhiri, A., N. Darwish, and N. Hilal, Membrane distillation: A comprehensive review. *Desalination*, 2012. **287**: pp. 2–18.
35. Cassard, H.M. and H.G. Park, How to select the optimal membrane distillation system for industrial applications. *Journal of Membrane Science*, 2018. **565**: pp. 402–410.
36. Law, R., A. Harvey, and D. Reay, A knowledge-based system for low-grade waste heat recovery in the process industries. *Applied Thermal Engineering*, 2016. **94**: pp. 590–599.
37. Ma, Q., A. Ahmadi, and C. Cabassud, Direct integration of a vacuum membrane distillation module within a solar collector for small-scale units adapted to seawater desalination in remote places: Design, modeling & evaluation of a flat-plate equipment. *Journal of Membrane Science*, 2018. **564**: pp. 617–633.
38. Gil, J.D. et al., A feedback control system with reference governor for a solar membrane distillation pilot facility. *Renewable Energy*, 2018. **120**: pp. 536–549.
39. Martinez, L. and J.M. Rodriguez-Maroto, Effects of membrane and module design improvements on flux in direct contact membrane distillation. *Desalination*, 2007. **205**(1–3): pp. 97–103.
40. Rom, A., W. Wukovits, and F. Anton, Development of a vacuum membrane distillation unit operation: From experimental data to a simulation model. *Chemical Engineering and Processing*, 2014. **86**: pp. 90–95.
41. Rao, G., S.R. Hiibel, and A.E. Childress, Simplified flux prediction in direct-contact membrane distillation using a membrane structural parameter. *Desalination*, 2014. **351**: pp. 151–162.

42. Kong, W. et al., A modified dusty gas model in the form of a Fick's model for the prediction of multicomponent mass transport in a solid oxide fuel cell anode. *Journal of Power Sources*, 2012. **206**: pp. 171–178.
43. Gao, F.X. et al., Compressible gases transport through porous membrane: A modified dusty gas model. *Journal of Membrane Science*, 2011. **379**(1–2): pp. 200–206.
44. Martinez, L. et al., Estimation of vapor transfer coefficient of hydrophobic porous membranes for applications in membrane distillation. *Separation and Purification Technology*, 2003. **33**(1): pp. 45–55.
45. Criscuoli, A., M.C. Carnevale, and E. Drioli, Modeling the performance of flat and capillary membrane modules in vacuum membrane distillation. *Journal of Membrane Science*, 2013. **447**: pp. 369–375.
46. Al-Khatib, A., *An Experimental Comparison of Performance between Flat Sheet and Hollow Fiber Membrane Modules in Direct Contact Membrane Distillation System for Desalination of Seawater*. 2016, Qatar University (Qatar): Ann Arbor, MI. p. 238.
47. Wan, C.F. et al., Design and fabrication of hollow fiber membrane modules. *Journal of Membrane Science*, 2017. **538**: pp. 96–107.
48. Yang, X., R. Wang, and A.G. Fane, Novel designs for improving the performance of hollow fiber membrane distillation modules. *Journal of Membrane Science*, 2011. **384**(1): pp. 52–62.
49. Teoh, M.M., S. Bonyadi, and T.-S. Chung, Investigation of different hollow fiber module designs for flux enhancement in the membrane distillation process. *Journal of Membrane Science*, 2008. **311**(1–2): pp. 371–379.
50. Winter, D., J. Koschikowski, and M. Wieghaus, Desalination using membrane distillation: Experimental studies on full scale spiral wound modules. *Journal of Membrane Science*, 2011. **375**(1): pp. 104–112.
51. Winter, D., J. Koschikowski, and S. Ripperger, Desalination using membrane distillation: Flux enhancement by feed water deaeration on spiral-wound modules. *Journal of Membrane Science*, 2012. *423–424*: pp. 215–224.
52. Ruiz-Aguirre, A., D.C. Alarcon-Padilla, and G. Zaragoza, Productivity analysis of two spiral-wound membrane distillation prototypes coupled with solar energy. *Desalination and Water Treatment*, 2015. **55**(10): pp. 2777–2785.
53. Li, J.-M. et al., Microporous polypropylene and polyethylene hollow fiber membranes. Part 3. Experimental studies on membrane distillation for desalination. *Desalination*, 2003. **155**(2): pp. 153–156.
54. Khayet, M., K. Khulbe, and T. Matsuura, Characterization of membranes for membrane distillation by atomic force microscopy and estimation of their water vapor transfer coefficients in vacuum membrane distillation process. *Journal of Membrane Science*, 2004. **238**(1–2): pp. 199–211.
55. Wang, X. et al., Feasibility research of potable water production via solar-heated hollow fiber membrane distillation system. *Desalination*, 2009. **247**(1): pp. 403–411.
56. Guijt, C.M. et al., Air gap membrane distillation - 2. Model validation and hollow fibre module performance analysis. *Separation and Purification Technology*, 2005. **43**(3): pp. 245–255.
57. Thomas, N. et al., Membrane distillation research & implementation: Lessons from the past five decades. *Separation and Purification Technology*, 2017. **189**: pp. 108–127.

58. Yazgan-Birgi, P., M.I. Hassan Ali, and H.A. Arafat, Comparative performance assessment of flat sheet and hollow fiber DCMD processes using CFD modeling. *Separation and Purification Technology*, 2019. **212**: pp. 709–722.
59. Zhongwei, D., Study on the effect of flow maldistribution on the performance of the hollow fiber modules used in membrane distillation. *Journal of Membrane Science*, 2003. **215**(1–2): pp. 11–23.
60. Ali, A., P. Aimar, and E. Drioli, Effect of module design and flow patterns on performance of membrane distillation process. *Chemical Engineering Journal*, 2015. **277**: pp. 368–377.
61. García-Fernández, L., C. García-Payo, and M. Khayet, Hollow fiber membranes with different external corrugated surfaces for desalination by membrane distillation. *Applied Surface Science*, 2017. **416**: pp. 932–946.
62. Aryapratama, R. et al., Performance evaluation of hollow fiber air gap membrane distillation module with multiple cooling channels. *Desalination*, 2016. **385**: pp. 58–68.
63. Mejia Mendez, D.L. et al., Membrane distillation (MD) processes for water desalination applications: Can dense selfstanding membranes compete with microporous hydrophobic materials? *Chemical Engineering Science*, 2018. **188**: pp. 84–96.
64. Singh, D. et al., Novel cylindrical cross-flow hollow fiber membrane module for direct contact membrane distillation-based desalination. *Journal of Membrane Science*, 2018. **545**: pp. 312–322.
65. Wang, P. and T.S. Chung, Recent advances in membrane distillation processes: Membrane development, configuration design and application exploring. *Journal of Membrane Science*, 2015. **474**: pp. 39–56.
66. Swaminathan, J. et al., Energy efficiency of permeate gap and novel conductive gap membrane distillation. *Journal of Membrane Science*, 2016. **502**: pp. 171–178.
67. Francis, L. et al., Material gap membrane distillation: A new design for water vapor flux enhancement. *Journal of Membrane Science*, 2013. **448**: pp. 240–247.
68. Alsaati, A. and A.M. Marconnet, Energy efficient membrane distillation through localized heating. *Desalination*, 2018. **442**: pp. 99–107.
69. Huang, L. et al., Water desalination under one sun using graphene-based material modified PTFE membrane. *Desalination*, 2018. **442**: pp. 1–7.
70. Cipollina, A. et al., CFD simulation of a membrane distillation module channel. *Desalination and Water Treatment*, 2009. **6**(1–3): pp. 177–183.
71. Cipollina, A., G. Micale, and L. Rizzuti, Membrane distillation heat transfer enhancement by CFD analysis of internal module geometry. *Desalination and Water Treatment*, 2011. **25**(1–3): pp. 195–209.
72. Yazgan-Birgi, P., M.I. Hassan Ali, and H.A. Arafat, Estimation of liquid entry pressure in hydrophobic membranes using CFD tools. *Journal of Membrane Science*, 2018. **552**: pp. 68–76.
73. Rezakazemi, M., CFD simulation of seawater purification using direct contact membrane desalination (DCMD) system. *Desalination*, 2018. **443**: pp. 323–332.
74. Amigo, J., R. Urtubia, and F. Suárez, Exploring the interactions between hydrodynamics and fouling in membrane distillation systems: A multiscale approach using CFD. *Desalination*, 2018. **444**: pp. 63–74.
75. Katsandri, A., A theoretical analysis of a spacer filled flat plate membrane distillation modules using CFD: Part II: Temperature polarisation analysis. *Desalination*, 2017. **408**: pp. 166–180.

76. Katsandri, A., A theoretical analysis of a spacer filled flat plate membrane distillation modules using CFD: Part I: velocity and shear stress analysis. *Desalination*, 2017. **408**: pp. 145–165.
77. Chang, H., C.-D. Ho, and J.-A. Hsu, Analysis of heat transfer coefficients in direct contact membrane distillation modules using CFD simulation. 淡江理工學刊, 2016. **19**(2): pp. 197–206.
78. Hayer, H., O. Bakhtiari, and T. Mohammadi, Analysis of heat and mass transfer in vacuum membrane distillation for water desalination using computational fluid dynamics (CFD). *Desalination and Water Treatment*, 2015. **55**(1): pp. 39–52.
79. Jafari, P. and M. Keshavarz Moraveji, Application of generic cubic equations of state in the CFD simulation of the sweeping gas polytetrafluoroethylene (PTFE) membrane distillation. *Desalination and Water Treatment*, 2014. **57**(4): pp. 1647–1658.
80. Shakaib, M. et al., A CFD study of heat transfer through spacer channels of membrane distillation modules. *Desalination and Water Treatment*, 2013. **51**(16–18): pp. 3662–3674.
81. Al-Sharif, S. et al., Modelling flow and heat transfer in spacer-filled membrane distillation channels using open source CFD code. *Desalination*, 2013. **311**: pp. 103–112.
82. Yang, X. et al., Analysis of the effect of turbulence promoters in hollow fiber membrane distillation modules by computational fluid dynamic (CFD) simulations. *Journal of Membrane Science*, 2012. *415–416*: pp. 758–769.
83. Yang, X. et al., Optimization of microstructured hollow fiber design for membrane distillation applications using CFD modeling. *Journal of Membrane Science*, 2012. *421–422*: pp. 258–270.
84. Shakaib, M. et al., A CFD study on the effect of spacer orientation on temperature polarization in membrane distillation modules. *Desalination*, 2012. **284**: pp. 332–340.
85. Soukane, S., J.-G. Lee, and N. Ghaffour, Direct contact membrane distillation module scale-up calculations: Choosing between convective and conjugate approaches. *Separation and Purification Technology*, 2019. **209**: pp. 279–292.
86. Mohammadi Ghaleni, M., M. Bavarian, and S. Nejati, Model-guided design of high-performance membrane distillation modules for water desalination. *Journal of Membrane Science*, 2018. **563**: pp. 794–803.

5

Membrane Distillation Performance Analysis

5.1 Introduction

Membrane distillation (MD) is a thermally driven membrane separation process where permeate flux is affected by a number of factors. Some factors are related to process conditions while others are related to membrane properties and enclosing module design in the broadest sense. The performance analysis of membrane distillation rests on two important indicators: permeate flux and energy efficiency. Much of the theoretical concepts that have been used and continue to be used for MD performance analysis have been put forward nearly 3 decades ago [1] and continue to be relevant. The general equation for the permeate flux is shown in equation (5.1):

$$J = C_m\left(P_{fm} - P_{pm}\right) \tag{5.1}$$

where J is the permeate flux, C_m is the hydrophobic membrane coefficient that is reported almost independent of process conditions in most cases [2–4], and P_{fm} and P_{pm} are the vapor pressure values on the membrane surface at feed and permeate side respectively. While the vapor pressure is traditionally estimated from Antoine equation-based correlations for pure water or very dilute solutions, it can be better estimated under wide ranges of conditions using the latest available correlations for seawater feed from the Massachusetts Institute of Technology (MIT) group led by Lienhard [5]. These correlations are updates of previously published ones by members of the MIT group [6].

It is important to bear in mind that in equation (5.1), the vapor pressures that constitute the MD driving force are determined at the membrane surface where the temperature simply cannot be measured easily or accurately. Hence, we rely on some modeling concepts to obtain the best estimates of such membrane surface temperatures.

The classic diagram representation of the driving force for membrane distillation is depicted in Figure 5.1 where a general membrane distillation process and important process variables are shown in three distinct zones, irrespective of the configurations described in the literature. The vapor flux starts at the hot feed side of the MD module exactly at the interface between the membrane surface and the hot side boundary layer and proceeds through the microporous hydrophobic membrane right up to the interface between membrane surface in the cold side (permeate side) and the cold side boundary layer where the collection method of the permeate depends on the module configuration described in Chapter 2. In Figure 5.1 we observe a number of important MD process variables: T_{fb} is the hot feed "bulk" temperature that can be easily measured and controlled at the module feed (by means of a sensor linked to a display of some sort). T_{fm} is the membrane surface temperature at the feed side that cannot be easily measured but can be estimated by means of suitable model solutions (described in Chapter 8). We see the first temperature drop ΔT_1 in zone 1 determined by the extent of the boundary layer (also known as the thermal boundary layer) adjacent to the membrane surface on the feed side. This phenomenon is known as the temperature polarization, covered

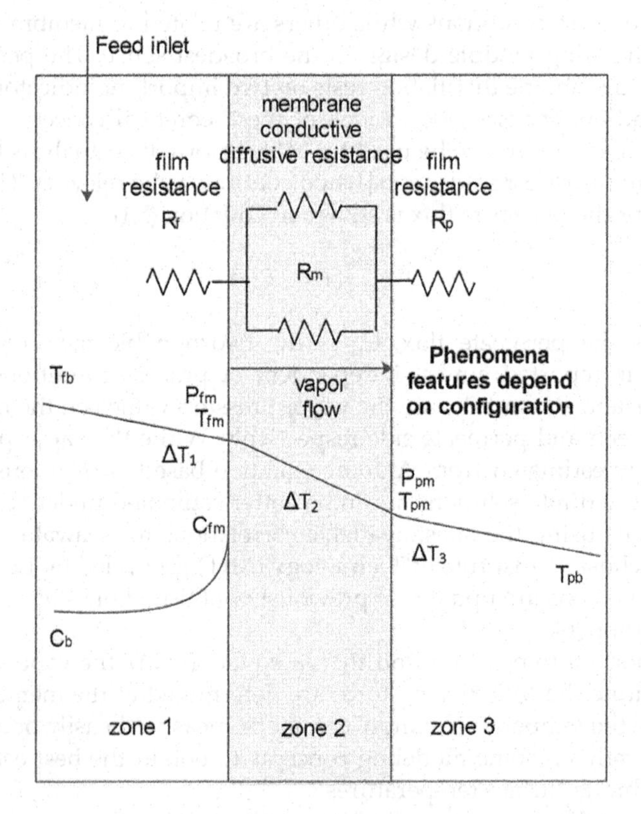

FIGURE 5.1
3-zones representation of driving force and resistances in membrane distillation of any configuration.

in Chapter 8. In Figure 5.1, the thermal boundary layer resistance R_f is also equal to $1/h_f$ where h_f is the feed side heat transfer coefficient. It is at this point where the "theoretical" feed vapor pressure P_{fm} on the membrane surface hot side is estimated. The system temperature trajectory proceeds through the porous membrane where it is subjected to a heat transfer resistance, namely the membrane thermal conductivity. A further temperature drop, ΔT_2 is observed in zone 2. At this point it is essential to point out that the porous membrane thermal conductivity is not the value of the "solid" membrane matrix but really it is the "effective thermal conductivity" of membrane where we have to consider the material of the membrane as well as the gas trapped in the membrane pores (air, vapor). The membrane effective thermal conductivity is estimated from literature models that combine in "some empirical" ways the solid thermal conductivity, the gas/vapor thermal conductivity and the membrane matrix porosity [3]. This important topic is covered further in Chapter 8. The thermal resistance of the membrane that leads to a temperature gradient within is also accompanied by a "mass transfer resistance" that slows the vapor transport within the membrane matrix due to the membrane porous nature described "theoretically" using the Dusty Gas Model [3,4,7–23]. We see from Figure 5.1 that the membrane resistance has two hypothetical contributions: thermal and mass transfer resistances. The temperature trajectory reaches the membrane surface in the permeate side (cold side) where it is denoted T_{pm}. At this point on the membrane surface, the "theoretical" permeate vapor pressure P_{pm} is estimated using temperature T_{pm} that is also estimated using modeling concepts covered in Chapter 8. Past the membrane surface on the permeate (cold) side (zone 3 in Figure 5.1), the temperature trajectory becomes a little complicated by the fact that a number of MD module configurations are possible. In addition, the notion of boundary layer, temperature polarization and permeate side resistance becomes conditional to MD system configuration design. Let us take a few simple cases to illustrate this situation:

- In direct contact membrane distillation (DCMD), as shown in Figure 5.1, we see in zone 3 that there is a thermal boundary layer adjacent to the membrane in the permeate side contributing further to the temperature polarization, leading to a further temperature difference ΔT_3 in zone 3 that represents the difference between the permeate bulk temperature T_{pb} and the membrane surface temperature T_{pm} in the permeate side. This is similar to the hot feed side case, and the permeate side resistance R_p can be assimilated to $1/h_p$ where h_p is the convective heat transfer coefficient obtained in the permeate side by means of a classical convective heat transfer approach using suitable Nusselt number correlations that are abundant in the literature, provided a careful selection process is adopted [20,24]. This is covered in detail in Chapter 8.

- In the case of an air gap membrane distillation (AGMD) configuration, there is a "stagnant mass of air" between the membrane permeate side and the condensation surface where distillate is collected, in addition to the potential air inside the membrane pores. Permeate vapor is transported by diffusion [25] through the stagnant layer of air, then through the condensate layer of condensed permeate, to the condensation surface [26]. In essence, for the AGMD case, the permeate side resistance shown in Figure 5.1, R_p is made up of a combination of the resistance through the air gap R_{ag}, plus the resistance through the condensation film R_{cf} and the resistance due to the thickness of the condensation plate R_{cp} enclosing the flowing coolant. The concept of resistances in AGMD is depicted in Figure 5.2. Clearly, the temperature polarization in AGMD is more complex than in the case of DCMD as several parameters (various thicknesses depicted in Figure 5.2 as δ with appropriate subscript, such as the stagnant air gap, condensation film and condensation plate) come into play.
- In the case of vacuum membrane distillation (VMD) and sweeping gas membrane distillation (SGMD), the resistances in the permeate side can be neglected since vacuum and flowing gas (sweeping) do not offer "significant" resistance to permeate vapor emerging from the membrane permeate side. This is considered as an advantage for these configurations when it comes to comparing flux performance for identical membrane characteristics and feed operational conditions.

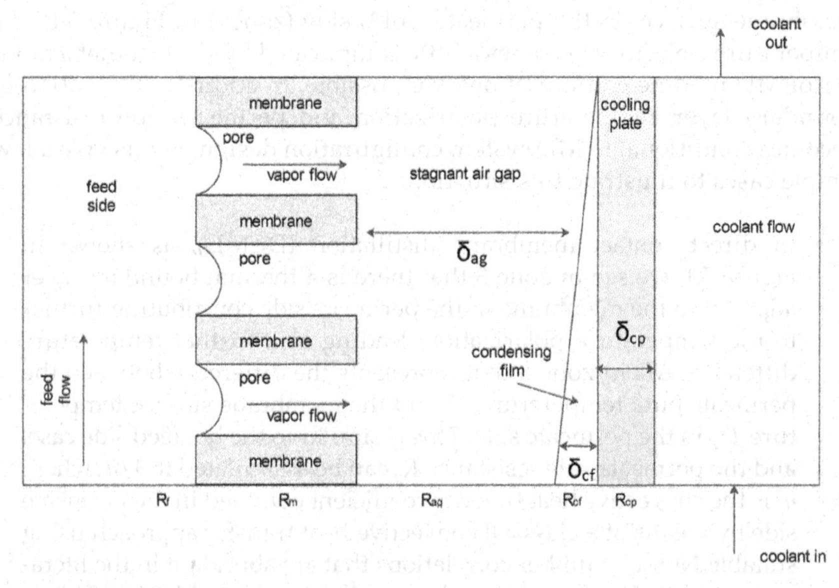

FIGURE 5.2
Various transport resistances in AGMD.

In equation (5.1), C_m, the membrane coefficient can be determined experimentally [3,4] or estimated using a variety of conceptual mass transfer models that require as input the membrane properties and operating temperature [20,27]. In either approach, the numerical values of such properties are "specific" to the particular type of membranes used. However, it is common practice to use experimental flux data to estimate the membrane coefficient C_m due to uncertainty in adopting a purely modeling approach with assumed membrane properties or limited membrane morphological properties. It has to be said that with currently used MD membranes, whether procured commercially or fabricated in the laboratory according to methods covered in Chapter 3 (sintering, stretching, trach-etching and phase inversion [28]), the morphological properties such as pore size distribution and porosity will almost invariably vary from batch to batch, albeit on a small scale, making exact replication almost impossible. This statement simply means that published membrane morphology data should be considered indicative and applicable to the case reported and not general for the types of membranes under consideration.

The discussion on the three zones depicted in Figure 5.1 provides a clue on how flux performance in membrane distillation can be analyzed and potentially improved. The major factors that can increase the flux are shown in Table 5.1. Unfortunately, the literature does not have an organized review where flux performance is reported according to the major factors shown in Table 5.1. Factors affecting flux can be related to membrane properties, operational conditions, feed concentration and module design as reported in recent review articles [29–31]. Much of the flux data are scattered, incomplete or have module configuration, membrane material, membrane morphology, operating conditions, feed salinity, mixed together [26,29,31], so that a full and deep analysis based on criteria in Table 5.1 is difficult to accomplish on small scattered samples of data. In this respect, the nearest review that is purely qualitative on factors affecting MD flux is that of El-Bourawi et al. [32]. However, a recent paper by Venneste et al. [33] reported an interesting work where authors analyzed flux and energy efficiency data from 17 MD membranes (commercial brands of membranes with materials polypropylene, PVDF and PTFE) of various thicknesses, pore sizes and porosity. The flux data were obtained at a feed temperature of 60°C and permeate temperature of 20°C. The flowrates of both feed and permeate were 1.6 L min^{-1} and the feed salinity was 1 g L^{-1} (1000 ppm). The use of a single set of temperatures (feed/permeate), single flowrate and a single fed salinity for all membranes made the study consistent and quite valuable and enabled the authors to put forward a new MD performance equation without prior knowledge of the membrane characteristics such as porosity, thickness and thermal conductivity [33].

Nevertheless, an attempt will be made to cover important performance affecting factors in order to draw some conclusions and pave the way to the following chapters that cover performance deterioration, improvements and modeling of performance.

TABLE 5.1
Major Factors That Increase the Flux

Zone in MD Configuration	Factors That Increase the Flux	Note
1	Increase in the feed temperature T_{fb}	The membrane surface temperature T_{fm} also increases. Antoine's vapor pressure is an exponential function of temperature
	Decrease in the permeate temperature T_{pb}	The membrane surface permeate side T_{pm} also decreases. This decreases the vapor pressure, permeate side, and increases the flux driving force $(P_{fm} - P_{pm})$
2	Decrease in conductive heat loss and decrease in transport resistance	Membrane material selection Membrane porosity increase Membrane pore size increase Degassing membrane pores Select VMD configuration
	Reduce membrane thickness	This is a trade-off between increased flux and increased heat loss by conduction. Compromise: thin hydrophobic layer for high flux and thicker support layer with larger pores to reduce heat loss
3	Decrease in the permeate temperature T_{pb}	The membrane surface permeate side T_{pm} also decreases. This decreases the vapor pressure, permeate side, and increases the flux driving force $(P_{fm} - P_{pm})$
	Use VMD or SGMD	VMD eliminates the resistance in zone 3 completely SGMD diminishes resistance to negligible level in zone 3

5.2 Distillate Flux Performance

5.2.1 Flat Sheet Membranes

5.2.1.1 Effect of Membrane Properties: Material, Thickness, Pore Size, Pore Size Distribution

It has been reported that the membrane distillation (MD) flux can be improved by reducing the membrane thickness δ and tortuosity τ or by increasing porosity ε and pore size r within the range of these proper for membrane distillation [8,27,34]. Eykens et al. investigated the effect of membrane thickness in a DCMD system and found that for pure water feed, the flux was relatively high for thinner membranes and that energy efficiency was unaffected by membrane thickness [35]. Eykens et al. also reported an optimal membrane thickness for concentrated feed ranging from 2 to 739 μm. This is a rather wide range, but authors provided a detailed table of useful relationships between salinity, temperature driving force (hot and cold side temperatures) and flux/energy efficiency. This table indicates an opposing trend. In other words, when the membrane thickness increases, the flux and energy efficiency decrease. It has to be noted that their temperature ranges were narrow and their results are therefore limited in scope. Other studies on the effect of membrane thickness indicated that the flux of thin membranes is more affected by salinity [36], optimal membrane thickness δ was within the range (30–60 μm) [12], while in another study, the optimal thickness δ was shown to be dependent on feed concentration and within the range (10–60 μm) [37]. Another study found that the optimal thickness δ depended on heat transfer in the channels, feed temperature and membrane permeability and was reported to be within a narrower range (10–20 μm) [38]. The above studies referred to the thickness of membranes without supporting material (for extra strength).

Ali et al. [39] carried out modeling simulations using literature data on membrane properties used to obtain membrane coefficients in order to extract flux, gained output ratio (GOR) and cost for direct contact membrane distillation (DCMD) and air gap membrane distillation (AGMD) configurations. They obtained, as expected, trends indicating the opposing effect between performance and membrane thickness. However, their simulation work showed that the cost could be much lower for AGMD compared to DCMD. This suggests that AGMD could be less sensitive to membrane properties compared to DCMD.

Swaminathan et al. carried out an interesting piece of work in which they presented a comprehensive analytical framework for a single stage MD system where the membrane thickness and system size were parameters in a flux and energy efficiency optimization context [40]. Their analysis was rather laborious but provided a useful insight into membrane thickness and system size to target optimum performance. They also confirmed previously reported findings related to a so-called optimum membrane thickness: when the membrane thickness is decreased, the vapor flux initially increases when treating salty water, obviously due to higher permeability. However, the flux starts declining due to conduction losses when thickness goes below a "certain" optimal value.

5.2.1.2 Effect of Temperature

The effect of temperature on MD flux performance should be scrutinized not only by the feed temperature alone but also including the permeate side temperature such that a meaningful insight into the temperature effect on the MD flux driving force can be gained. The flux equation depicted by equation (5.1) clearly shows the indirect temperature dependency of the vapor pressure difference driving force. In addition, temperature has a milder effect on diffusive vapor transport inside the porous membrane. The nature of temperature dependency on water vapor diffusive transport within the membrane porous matrix can be found in the membrane coefficient and this in turn has mathematical expressions that relate to the type of transport assumed (Knudsen, molecular, Poiseuille or a combination of these) [20,26,27,29,41,42].

All reported work on the effect of temperature on membrane distillation flux suggest that an increase in feed temperature coupled with a decrease in temperature in the permeate side lead to higher permeate flux. While this important observation can tempt one to adopt as wide a temperature difference as practical (for instance feed temperature ranging 70°C–80°C and permeate temperature ranging 15°C–25°C), caution must be exercised as the system energy efficiency and risk of membrane scaling can deteriorate, as will be shown in subsequent chapters on performance deterioration in membrane distillation. Indeed, temperature alone is not a good indicator of membrane distillation flux performance. Other factors such as feed/permeate flowrates, feed salinity and membrane pore size (and porosity) also play a crucial role [43].

5.2.1.3 Effect of Flowrates and Feed Recirculation

The feed and permeate flowrate affect the convective heat transport near the membrane surface (both hot and cold side), as can be deduced from Figure 5.1 (film resistances R_f and R_p). The flowrate determines the regime of flow inside the membrane distillation (MD) module (laminar, turbulent or transitional) and consequently the flux performance. This is particularly important in modeling work and in heat transfer calculations in MD [3].

The feed flowrate can affect the extent of the boundary layer adjacent to the membrane surface and therefore can theoretically affect the temperature and concentration polarization phenomena (Figure 5.1). However, the effect of flowrate was reported to be significant only at temperatures higher than about 40°C [44]. As the flowrate increases, the flux appears to increase but levels off beyond a certain threshold, indicating an almost asymptotic behavior [45,46].

In the permeate side, the effect of flowrate for direct contact membrane distillation (DCMD) is less pronounced than in the feed side but still has a significance on the convective heat transfer coefficient (Figure 5.1). Thus, increasing the flowrate has a positive but limited effect. For other configurations such as sweeping gas membrane distillation (SGMD), the flowrate of sweeping gas has a net effect of reducing the extent of mass transfer resistance to vapor transport [47,48]. The vacuum and air gap membrane distillation configurations have no relevance to flowrate in the permeate side.

The bulk of studies in membrane distillation were conducted on the basis of single pass feed, thus giving low distillate recovery rates, typically less than 10% [49,50] compared to reverse osmosis where permeate recovery can be quite high, ranging 50%–84% [51]. In membrane distillation, the distillate (or permeate) recovery rate (RR) is defined as:

$$RR = \frac{M_{permeate}}{M_{feed}} \tag{5.2}$$

$$RR_{per\ pass} = \frac{\dot{m}_p}{\dot{m}_f} \tag{5.3}$$

where M and $\dot{m}$ are mass and mass flowrate respectively. The subscripts denote the respective nature of the stream (feed or permeate).

In order to improve the permeate recovery rate RR in currently reported single stage membrane distillation work, recent work suggested the following optional operations [50]:

- Batch recirculation
- Semi-batch recirculation
- Continuous recirculation
- Continuous multi-stage recirculation

Obviously, these operations will have implications such as variable feed salinity (increasing) and more complicated module design, adding perhaps new challenges and opportunities for membrane distillation desalination.

On the same subject of attempting to improve permeate recovery rate RR, Lokare et al. [51] proposed a simulation based approach using the recovery equations below that exploit the system evaporation efficiency η:

$$mx = \frac{\eta\left[mc_pT_{in} - m(1-x)c_pT_{out}\right]}{100L} \tag{5.4}$$

$$x = \frac{T_{in} - T_{out}}{\dfrac{100L}{\eta c_p} - T_{out}} \tag{5.5}$$

where m is the feed flowrate, x is the fraction of feed that is recovered on the permeate side, η is the evaporation efficiency (%), c_p is the specific heat capacity, L is the latent heat of evaporation, T_{in} and T_{out} are the feed inlet and outlet temperatures respectively.

The evaporation efficiency η is defined by equation (5.6) [52–54]:

$$\eta = \frac{JL}{Q_{total}} = \frac{JL}{JL + Q_{cond}} = \frac{JL}{c_p\dot{m}\Delta T_l} \tag{5.6}$$

where η is the evaporation efficiency, J is the flux, L is the latent heat of evaporation, Q_{total} is the total heat flux transferred across the membrane, Q_{cond} is the heat loss through conduction, c_p is the specific heat capacity of water, $\dot{m}$ is the liquid mass flowrate and ΔT_1 is the temperature difference between T_{in} and T_{out} (feed inlet and outlet temperatures).

From equations (5.4) through (5.6), it can be seen that the maximum amount of permeate that can be recovered from the feed in a single pass can be estimated if the evaporation efficiency η of a system is known [51]. η can be obtained from equation (5.6) using experimentally or computationally determined distillate flux J. Through a series of simulations involving feed recirculation, Lokare et al. showed that the recovery rate increases are also accompanied by a significant increase in energy consumption. They reported that an increase in water recovery in a direct contact membrane distillation module from 10% to 50% for a feed solution of 100,000 ppm NaCl would lead to an increase in the required recycle ratio by 633% (from 3 to 22) with a corresponding increase in thermal energy needed to reheat the recycle stream by 556% (an increase from 39 to 256 kWh/m^3 of feed). They also indicated that while the electrical energy required for feed recirculation represents only a few percent of thermal energy requirements, it may well constitute a significant factor when considering the overall life cycle impacts of the membrane distillation desalination process [51].

5.2.1.4 Effect of Turbulence Promoters (Spacers)

Enhancing turbulent flow in membrane distillation can have a beneficial effect in terms of reducing the negative impact of the boundary layer and hence reduce the temperature and concentration polarization phenomena. The obvious means would be to increase the flowrate. However, it is not always practical to do so and hence the use of turbulence promoters (also called in the industry "spacers"). These are essentially "static mixers" inserted in the flat sheet module in the hot or cold side (or both). The idea of using spacers in membrane distillation is borrowed from reverse osmosis, and a great deal of studies have been done in membrane distillation using spacers [17,55–76] involving laboratory as well as simulation studies.

One of the most comprehensive recent studies on the effect of spacers on MD flux is that of Kim et al. [76]. The reported permeate flux enhancement by the spacers ranging 7%–19% for the spacer-filled permeate channels while the improvement in flux ranged 21%–33% for the spacer-filled feed channels. They also reported that the influence of spacers on flux improvement became more pronounced at higher temperatures due to higher temperature polarization. In their study, Kim et al. reported that the maximum flux enhancement (approximately 43%) over the empty channels was achieved using the thinnest and densest spacer with a hydrodynamic angle of 90°, adjacent to both membrane surfaces [76]. Earlier studies on the effect of spacers had a useful contribution toward showing the benefits of such turbulence promoters [3,57,58,77] and also improving modeling studies since these studies provided "corrections" to traditional Nusselt number correlations [3] for the estimation of thermal coefficients in convective heat transfer in membrane distillation.

5.2.1.5 Effect of Flow Direction (Counter-Current vs Co-current)

Although direct contact membrane distillation (DCMD) modules can be operated in counter or co-current operations, nearly the bulk of studies were conducted in counter-current flow mode owing to its superiority in terms of temperature difference between the feed and permeate streams. Some studies were conducted in which these modes of operations were compared [78,79]. The overall flux appeared to be always higher in counter-current mode and the difference can be slight [46] to moderate [78]. The lack of interest in co-current flow is also motivated by disadvantages when it comes to heat recovery for multistage modules. One can say that co-current flow is so far only for academic studies and for comparison purpose to justify the "obvious" choice of counter-current.

5.2.1.6 Effect of Feed Concentration

In membrane distillation, the permeate flux is adequately described by equation (5.1) where one of the important factors is the vapor pressure difference. The vapor pressure of an aqueous solution is known to be affected by the non-volatile solute concentration for quite some time. Fortunately a great deal of efforts were deployed to study experimentally seawater properties, and virtually all physical properties of seawater of interest to membrane distillation desalination have been collated in the form of correlations by the Lienhard research group at Massachusetts Institute of Technology (MIT) [5,6,80]. Simple calculations show that the vapor pressure of a saline solution is less than that of pure water and that it decreases slightly as the salt content increases. How does the feed salt concentration affect the flux performance in membrane distillation? There have been numerous studies on this topic and the consensus is that an increase in feed salinity always leads to a decrease in permeate flux. The review paper of El-Bourawi et al. [32] reported some 25 publications that indicated a flux reduction with an increase in feed non-volatile concentration. That was in 2006. Since then many more papers confirmed the findings. Now, the question might be to what extent is the permeate flux decreased by an increase in salt content in the feed? One of the earliest quantitative estimations is that reported by Martinez and Rodriguez-Maroto [81] where an increase salt molar concentration from 0 to 4 yielded an approximate 15% decrease in permeate flux (at 36°C/16°C feed/permeate temperatures respectively). The flux reduction was attributed solely to vapor pressure depression.

More recent work on the effect of feed concentration is that of He et al. [46], which showed that the permeate flux decreased by a little over 20% when the feed was changed from pure water to seawater (or synthetic seawater).

5.2.2 Hollow Fiber (Capillary Membranes)

Section 5.2.1 was devoted to flat sheet membranes in corresponding modules. What about the flux performance using hollow fiber (or capillary) membranes? The factors that affect flux in flat sheet membranes also do so in hollow fiber membranes generally speaking. However, the major difference is not so much on the membranes alone, but in the module design and how the flow regime differs inside the module. Indeed, for the case of a hollow fiber membrane module, assuming the feed is allocated to the shell side and the permeate to the hollow fiber tube side, we notice immediately the potential for flow maldistribution. This was studied by Zhongwei et al. [82] who reported the non-uniformity of fiber packing as a possible source for flow maldistribution. Hollow fiber membrane modules for general industrial applications are known to have productivity issues [83] and much of these issues were attributed to poor module design in terms of hydrodynamics performance.

The distillate flux performance in hollow fiber membrane modules has been reported to be lower [84–90] than in flat sheet membrane modules [3,4,91–97]. The reasons for lower fluxes in hollow fiber membrane modules without internal flow enhancements were attributed to poor hydrodynamics and lower turbulence that in turn could lead to lower convective thermal coefficients.

Qu et al. [98] and Al-Khatib [90] reported thermal coefficients that were thought to be lower in hollow fiber membrane modules than in the case of flat sheet modules, confirming that the flow regime in hollow fiber modules could be one of the weaknesses that needed addressing to improve flux performance. One of the very few studies on a comparison in flux performance between flat sheet and hollow fiber modules under very similar conditions is that of Al-Khatib [90]. The comparative study of Singh and Sirkar [99] involved membranes with different pore sizes and porosities, making the flux comparison somewhat indicative. In some cases, the difference in flux performance and the convective heat transfer coefficients were different by 10 fold and were explained in terms of differing flow regimes [90]. The recommendations put forward to improve flux performance (and any other derived performance such as energy efficiency) included more research on promoting better flow regimes and enhanced turbulence in the shell side of hollow fiber membrane modules. This recommendation was in fact already initiated in a number of studies at the Singapore Membrane Technology Centre [100,101].

5.2.3 Multistage MD Systems and Novel Module Design

Chung et al. [102] presented the performance of a multistage vacuum membrane distillation (MSVMD), which is thermodynamically similar to a multistage flash distillation (MSF) and has been evaluated for desalination, brine concentration and produced water reclamation applications. A wide range of solution concentrations were accurately modeled. Energy efficiency, gained output ratio (GOR), second law efficiency and the specific membrane area were used to quantify the performance of the system.

Alsaadi et al. [103] reported in a paper that the effect of temperature polarization was decoupled from the membrane mass transfer coefficient by preventing the liquid feed stream from contacting the membrane surface through the use of a novel custom-made vacuum MD (VMD) module design.

A new module design for membrane distillation, namely material gap membrane distillation (MGMD), for seawater desalination has been proposed and successfully tested by Francis et al. [104]. They claimed that an increase in the water vapor flux of about 200%–800% was observed by filling the gap with sand and deionised water.

Gilron et al. [105] proposed higher water recovery through cascades of cross flow DCMD modules. Depending on configuration (interstage heating or not) and temperature differences, recoveries of 60% and GOR values above 20 may be attainable in their design.

5.3 Energy Efficiency

In membrane distillation the efficient use of energy to produce distillate from low temperature energy sources (low-grade waste heat, renewable solar for instance) can be described in a number of ways according to the literature. Much is borrowed from traditional thermal desalination and reverse osmosis. For thermal desalination, the gained output ration (GOR) is often used to reflect the amount of steam consumed per amount of distillate produced. GOR can be obtained from equation (5.7):

$$GOR = \frac{\dot{m}_p \Delta H_v}{\dot{Q}_{input}} \tag{5.7}$$

where $\dot{m}_p$ is the permeate flowrate, ΔH_v is the latent heat of the permeating vapor (estimated at a low temperature, since low-grade heat is used) and $\dot{Q}_{input}$ is the heat flow input. For single stage membrane distillation, GOR is usually reported as a value less than 1 [106]. While for multiple stages, GOR can exceed the value of 1 [107]. Other expressions for GOR using the system thermal efficiency and have been reported [107–110]. These GOR expressions are best used in conjunction with numerical modeling of MD systems with an aim to improve thermal performance using various scenarios involving membrane properties, MD module design and operating conditions.

Energy efficiency and thermal energy efficiency in membrane distillation have different meanings. While the thermal energy efficiency relates to purely heat energy such as latent heat and conductive heat limited to the boundaries of the membrane system of analysis, energy efficiency in general includes other forms of energy such as electrical energy in addition to heat energy and extends well beyond the membrane system boundaries to include pumping, vacuum creating energy, refrigeration/cooling energy, etc. In addition, most literature only reports thermal energy efficiencies and ignores heat losses in pipes and tanks (the assumption being perfect insulation of these).

For the straightforward direct contact membrane distillation system (DCMD), the thermal efficiency without losses to the surrounding environment is described by equation (5.8) [111]:

$$\eta_{thermal} = \frac{J\Delta H_v}{J\Delta H_v + \frac{k_m}{\delta}\left(T_{mf} - T_{mp}\right)} \tag{5.8}$$

where J is the permeate flux, ΔH_v is the vapor latent heat, k_m is the thermal conductivity of the membrane, δ is the membrane thickness and T_{mf}, T_{mp} are the temperatures of the membrane surface at the feed and permeate side respectively.

Equation (5.8) provides useful clues on what affects the thermal efficiency in membrane distillation: the membrane properties (thickness and thermal conductivity) and the operating temperatures.

To gain a deeper insight into the effect of membrane property, feed and permeate temperatures, Deshmukh and Elimelech [112] expanded equation (5.8) further using a Taylor series approximation for the flux and used the membrane material thermal conductivity coupled with the porosity and thermal conductivity of the gas in the pores. With the modified equation (5.8), they produced plots of the membrane porosity versus thermal efficiency and membrane mean temperature versus thermal efficiency. The plots obtained were quite useful in seeing clearly trends: thermal efficiency increases with porosity and with mean membrane temperature. These findings were also confirmed by Zhang et al. in a review paper [111]. Zhang et al. reported that vacuum membrane distillation (VMD) had higher thermal efficiency than direct contact membrane distillation (DCMD) [111]. For instance, the thermal efficiency of the VMD process varied from 88.1% to 91.9% while the thermal efficiency for DCMD varied from 59.6% to 70.5% when the feed temperature was changed from 50°C to 85°C. However, as indicated previously, thermal efficiency is not a good indicator of overall energy efficiency. The seemingly high thermal efficiency of the VMD is somewhat diminished if the vacuum pump energy and the condenser energy are accounted for. Other negating factors include the risk of membrane wetting or damage due to the high-pressure difference in VMD. Zhang et al. [111] also compared air gap membrane distillation (AGMD) and DCMD in terms of thermal efficiency and found that the AGMD system had an exceptionally high thermal efficiency because of the low thermal conductivity of the air gap and that AGMD can prevent the conductive losses associated with DCMD. The AGMD was reported to have a relatively high heat efficiency of 0.70–0.98, but the introduction of the air gap and the cooling surface increases the complexity of the module and the mass transfer resistance, which ultimately may lead to a lower flux than that of DCMD under the same driving force. They emphasized the importance of heat recovery to enhance thermal efficiency but did not elaborate on the cost benefits of doing so.

Deshmukh and Elimelech [112] arrived at similar conclusions as Zhang et al. [111] and proposed an optimum membrane thickness of around 95 microns.

Elimelech and Phillip put a theoretical limit on the minimum energy requirements for desalination regardless of the technology used through simple thermodynamic calculations [113]. This theoretical limit stands at around 1.06 kWh/m^3 for a feed of 35,000 ppm salinity and 50% recovery, while an ideal "practical" limit was estimated as 1.56 kWh/m^3 of distillate for the best technology that can approach this energy consumption target. A well-designed reverse osmosis plant or pilot plant using the "gold standard" thin film composite semi permeable membrane can achieve an energy consumption of around 2 kWh/m^3, which is very close to the ideal "practical" lower limit on energy consumption. Criscuoli et al. [114] reported energy consumptions between 1.15 and 3.55 $kW/(kgh^{-1})$ for laboratory DCMD operating at 59°C/13.4°C feed/permeate temperatures, and VMD operating at 59°C feed temperature. However, the feed for these laboratory setups was distilled water not a saline solution, and the system was well insulated. The reason provided for using this type of feed was avoiding the effect of concentration on performance.

Khayet compiled reported literature GOR values and specific energy consumption values that varied widely [115]. He reported GOR values varying from 0.3 to 8.1 for a variety of MD systems using solar power. He also reported specific energy consumption of MD systems varying from 1.5 to 4580 kWh/m^3. The great scatter and wide variation in reported energy efficiency and consumption in membrane distillation was attributed to multiple factors that include lack of clarity on how these values were obtained and whether assumptions made were indeed valid. Also, the bulk of research was conducted on small-scale laboratory MD apparatus that do not represent the reality of pilot plant of industrial scale units where energy efficiency (as opposed to thermal efficiency) and energy consumption make more sense.

Al-Obaidani et al. [116] reported thermal efficiency and flux data for hollow fiber membrane direct contact modules for various salinity feeds and different membrane material (polypropylene PP, polyvinylidene fluoride PVDF and polytetrafluoroethylene PTFE), operating at 55°C/25°C (feed/distillate). The thermal efficiency declined from around 58% to 40% when the feed salinity increased from 35 to 335 g/L, due to a reduction in the heat of vaporization. The best reported thermal efficiencies for membrane materials were 65%, 58% and 52% for PP, PVDF and PTFE respectively when temperatures of 55°C/30°C were used [116].

Alklaibi and Lior [117] reported a comparative study of thermal energy efficiency between DCMD and AGMD. They reported that AGMD had consistently higher thermal efficiency compared to DCMD. However, their study provided an interesting trend in thermal efficiencies of AGMD and DCMD when the feed temperature was kept constant at 70°C and the permeate temperature was varied from 5°C to 45°C. They found that while the thermal efficiency of AGMD increased very slightly from 94% to 96% (only 2% variation), that of DCMD increased significantly from 83% to 93% (12% variation). This remarkable result shows the sensitivity of DCMD to the permeate temperature with respect to membrane heat losses. DCMD is well known to have the least thermal efficiency in all configurations reported in the literature [118].

Clearly, membrane distillation (MD) cannot compete with a well-designed reverse osmosis plant in terms of energy consumption, but the advantage of MD would be the utilization of low-grade waste heat at low temperatures (50°C–70°C) or solar energy through solar stills.

Applying thermodynamic laws with chemical thermodynamics of electrolytes and conducting an exergy analysis, Lienhard et al. [119] arrived at the staggering results shown in Table 5.2 (data extracted from [119]), showing the superior efficiencies of work-based desalination compared to thermal technologies including membrane distillation that has the lowest exergetic efficiency of 1%. The results in Table 5.1 are indicative and point out the low energy efficiency of thermally based desalination methods in general. The work of Lienhard et al. [119] suggests that the current status of scatted in energy efficiency, and energy consumption data in membrane distillation reported by Khayet [115] deserves a serious review to establish more reliable energy consumption and efficiency data if commercial deployment is to be a reality one day.

TABLE 5.2

Exergetic Efficiency of a Selection of Desalination Technologies

Desalination Technology	Minimum Energy (Efficiency η, %)
Reverse osmosis	32
Mechanical vapor compression	9
Multi-effect distillation	6
Membrane distillation	1

5.4 Distillate Quality

Membrane distillation for desalination has a theoretical salt rejection of 100% and ultra pure distillate. Karakulski et al. [120] reported a distillate quality of 0.8 μS/cm electrical conductivity corresponding to about 0.6 ppm TDS (total dissolved solids). However, the feed was "potable" water from a surface reservoir and the MD treatment was to make it suitable for industrial applications where purity is desirable. Schneider et al. [121] reported an even better quality of 0.4 μS/cm electrical conductivity for a feed of up 26% NaCl synthetic brine solution. However, they reported a drop in distillate flux from 9 kg/(m^2 h) at 0.05% NaCl feed to 6.3 kg/(m^2 h) at 26% NaCl.

Banat et al. published work on a solar-powered pilot plant spiral wound PTFE membrane distillation module. The PTFE membrane had average or nominal pores of 0.2 μm and a porosity of 80%. The total membrane area was about 10 m^2. With a feed of real untreated seawater from the Red Sea, they obtained a distillate quality of 20 to 250 μS/cm. It is surprising that most research output on membrane distillation flux omit to mention the distillate quality. Salt rejection seems to be the preferred option.

The quality of the permeate not only depends on the feed salinity but also on the duration of the experiments (for laboratory work) or operations (for pilot plant work). The decrease in permeate quality with time (often associated with a decline in permeate flux as well) is due to membrane wetting. The flux decline and quality of permeate deterioration will be covered further in Chapter 6 on fouling and scaling.

While the permeate quality can be very high and suitable for industrial use, it would be necessary to mineralize the product water if it were for human or agricultural use. This is a standard procedure in seawater desalination for human consumption [122].

5.5 Field Testing

There have been far fewer pilot plant studies compared to laboratory studies in membrane distillation. One of the early pilot plant studies on membrane distillation is that of Andersson et al. [123]. The paper was descriptive with few details on performance.

Koschikowski et al. [107] reported on a solar-powered membrane distillation unit capable of producing 120 to 160 L of water during a day in the summer. They used a collector area less than 6 m^2 without heat storage. The solar-powered MD unit was in a developmental phase and provided valuable dynamic performance data for a renewable energy desalination unit.

Song et al. [124] conducted one of the most detailed studies on a membrane distillation pilot plant over a period of 3 months. Their direct contact module had hollow fiber membranes. The runs employed hot brine at 64°C–93°C and distillate at 20°C–54°C. The hot brine was either city water, city water containing salt (3.5%, 6% and 10% solutions) or seawater. They achieved a permeate flux of 15 to 55 kg/(m^2 h) depending on the number of the modules used.

Meindersma et al. [125] described a proprietary air gap membrane distillation pilot unit bearing the trade mark Memstill®.

The process characteristics for the Memstill® process are [125]:

- Specific flux: Js = 1.5·10–10 m^3/m^2.s. Pa
- Heat energy: 80–240 MJ/m^3
- Production: 25–50 m^3/d.module
- Recovery: 50%

A recent pilot pant study conducted by Minier-Matar et al. [126] involved two separate membrane distillation units: vacuum multi-effect MD unit, with 4 effects and air gap MD unit, single effect. Both units were self-contained, highly mobile and had remote data acquisition and control. It seems there were some of the most advanced pilot plants to be reported in the literature. The vacuum unit was able to consistently generate high quality product water (TDS < 10 mg/L) over a 12-day period, while maintaining a stable flux of 4.5/(m^2.h). Due to the high salt concentrations (70,000 ppm), a 10% reduction in flux was observed when compared to the tap water baseline tests (5.0 L/(m^2.h)). In addition, the vacuum MD unit was able to process thermal reject brine (71,031 ppm) to produce excellent quality distillate (TDS of 6 ppm) and achieve a salt rejection of 99.99%. On the other hand, the air gap unit performed less well, only achieving 98.92% salt rejection with a thermal brine feed of 68,529 ppm and producing distillate with TDS of 1472 ppm. The best performing vacuum MD unit showed signs of moderate flux decline for extended period usage using thermal brine feed. However, the flux did not decline at all when the seawater feed was injected with antiscalant. The vacuum MD unit was able to operate steadily for 8 days with distillate flux of 4.8 L/(m^2.h) and recovery of 30%.

Gil et al. [127] presented an advanced feedback control system for a solar-powered membrane distillation in Spain. The work involved several types of membrane distillation systems (liquid gap, vacuum and air gap membrane distillation systems) and were powered by stationary flat plate collectors with a maximum power of 7 kW at 90°C. Given the transient nature of solar power, this study is deemed to be one of the first steps toward ensuring reliability of solar-powered membrane distillation in remote areas through control systems.

5.6 Membrane Distillation System Optimization

Cassard et al. [128] presented in a recent paper a simple yet comprehensive method for selection of the optimal membrane distillation design for any industrial process. This paper takes advantage of the wealth and volume of information published to date to formulate an effective way to select the membrane, module and operating conditions to achieve a target performance.

5.7 Concluding Remarks

Membrane distillation for desalination and other water treatment processes is a thermally based process that delivers a permeate flux that depends on so many factors that great caution must be exercised when comparing flux performance of various MD systems and hydrophobic membranes and associated energy consumption. There is so much scatter in the literature data that there is an urgent need for a comprehensive, systematic organized study to enable a fair and objective comparison of performance. The scatter in energy consumption makes it particularly difficult to rate MD against other desalination technologies on an objective basis. The thermal efficiency of MD, which is assumed under no heat losses to the environment, does not really provide a reliable measure like energy consumption where a more realistic estimate of total energy input is used. All contributing energy inputs should be used to determine energy consumption per m^3 (or ton) permeate produced, including any form of cooling below the ambient datum. If all these conditions were to be met, it would be found that MD, being a thermally based technology, would not be seen as an energy efficient champion, but rather as a technology that can exploit low carbon, low-cost energy sources to produce freshwater to remote communities or to produce ultra pure water when thermally integrated with processes that release significant amounts of low-grade heat. Indeed, there is merit for these cases, and efforts should be deployed on lowering costs of tapping into solar energy or recovering low-grade heat, rather than trying to push energy efficiency to expensive limits that only increase capital costs.

References

1. Schofield, R.W., A.G. Fane, and C.J.D. Fell, Gas and vapour transport through microporous membranes. II. Membrane distillation. *Journal of Membrane Science*, 1990. **53**(1): pp. 173–185.

2. Schofield, R.W. et al., Factors affecting flux in membrane distillation. *Desalination*, 1990. **77**(Supplement C): pp. 279–294.
3. Phattaranawik, J., R. Jiraratananon, and A.G. Fane, Heat transport and membrane distillation coefficients in direct contact membrane distillation. *Journal of Membrane Science*, 2003. **212**(1): pp. 177–193.
4. Gustafson, R.D., J.R. Murphy, and A. Achilli, A stepwise model of direct contact membrane distillation for application to large-scale systems: Experimental results and model predictions. *Desalination*, 2016. **378**: pp. 14–27.
5. Nayar, K.G. et al., Thermophysical properties of seawater: A review and new correlations that include pressure dependence. *Desalination*, 2016. **390**: pp. 1–24.
6. Sharqawy, M.H., J.H. Lienhard, and S.M. Zubair, Thermophysical properties of seawater: A review of existing correlations and data. *Desalination and Water Treatment*, 2010. **16**(1–3): pp. 354–380.
7. Mason E.A. and A.P. Malinauskas, *Gas Transport in Porous Media: The Dusty-Gas Model*. Chemical Engineering Monographs. 1983, New York: Elsevier.
8. Lawson, K.W., M.S. Hall, and D.R. Lloyd, Compaction of microporous membranes used in membrane distillation. I. Effect on gas permeability. *Journal of Membrane Science*, 1995. **101**(1): pp. 99–108.
9. Lawson, K.W. and D.R. Lloyd, Membrane distillation. I. Module design and performance evaluation using vacuum membrane distillation. *Journal of Membrane Science*, 1996. **120**(1): pp. 111–121.
10. Lawson, K.W. and D.R. Lloyd, Membrane distillation. II. Direct contact MD. *Journal of Membrane Science*, 1996. **120**(1): pp. 123–133.
11. Guijt, C.M. et al., Determination of membrane properties for use in the modelling of a membrane distillation module. *Desalination*, 2000. **132**(1–3): pp. 255–261.
12. Laganà, F., G. Barbieri, and E. Drioli, Direct contact membrane distillation: Modelling and concentration experiments. *Journal of Membrane Science*, 2000. **166**(1): pp. 1–11.
13. Phattaranawik, J., R. Jiraratananon, and A.G. Fane, Effect of pore size distribution and air flux on mass transport in direct contact membrane distillation. *Journal of Membrane Science*, 2003. **215**(1): pp. 75–85.
14. Imdakm, A., A Monte Carlo simulation model for membrane distillation processes: Direct contact (MD). *Journal of Membrane Science*, 2004. **237**(1–2): pp. 51–59.
15. Curcio, E. and E. Drioli, Membrane distillation and related operations – A review. *Separation and Purification Reviews*, 2005. **34**(1): pp. 35–86.
16. Imdakm, A.O. and T. Matsuura, Simulation of heat and mass transfer in direct contact membrane distillation (MD): The effect of membrane physical properties. *Journal of Membrane Science*, 2005. **262**(1): pp. 117–128.
17. Martinez, L. and J.M. Rodriguez-Maroto, Characterization of membrane distillation modules and analysis of mass flux enhancement by channel spacers. *Journal of Membrane Science*, 2006. **274**(1–2): pp. 123–137.
18. Imdakm, A.O., M. Khayet, and T. Matsuura, A Monte Carlo simulation model for vacuum membrane distillation process. *Journal of Membrane Science*, 2007. **306**(1–2): pp. 341–348.
19. Qi, B.W., B.A. Li, and S.C. Wang, Investigation of shell side heat transfer in crossflow designed vacuum membrane distillation module. *Industrial & Engineering Chemistry Research*, 2012. **51**(35): pp. 11463–11472.

20. Hitsov, I. et al., Modelling approaches in membrane distillation: A critical review. *Separation and Purification Technology*, 2015. **142**: pp. 48–64.
21. Lee, J.-G. et al., Performance modeling of direct contact membrane distillation (DCMD) seawater desalination process using a commercial composite membrane. *Journal of Membrane Science*, 2015. **478**: pp. 85–95.
22. Chang, H., C.-D. Ho, and J.-A. Hsu, Analysis of heat transfer coefficients in direct contact membrane distillation modules using CFD simulation. 淡江理工學刊, 2016. **19**(2): pp. 197–206.
23. Perfilov, V., V. Fila, and J. Sanchez Marcano, A general predictive model for sweeping gas membrane distillation. *Desalination*, 2018. **443**: pp. 285–306.
24. Gryta, M. and M. Tomaszewska, Heat transport in the membrane distillation process. *Journal of Membrane Science*, 1998. **144**(1): pp. 211–222.
25. Schofield, R.W., A.G. Fane, and C.J.D. Fell, Gas and vapour transport through microporous membranes. I. Knudsen-Poiseuille transition. *Journal of Membrane Science*, 1990. **53**(1): pp. 159–171.
26. Khayet, M., Membranes and theoretical modeling of membrane distillation: A review. *Advances in Colloid and Interface Science*, 2011. **164**(1–2): pp. 56–88.
27. Drioli, E., A. Ali, and F. Macedonio, Membrane distillation: Recent developments and perspectives. *Desalination*, 2015. **356**: pp. 56–84.
28. Mulder, M., *Basic Principles of Membrane Technology*. 1997, Dordrecht, the Netherlands: Kluwer Academic Publishers.
29. Alkhudhiri, A., N. Darwish, and N. Hilal, Membrane distillation: A comprehensive review. *Desalination*, 2012. **287**: pp. 2–18.
30. Camacho, L. et al., Advances in membrane distillation for water desalination and purification applications. *Water*, 2013. **5**(1): p. 94.
31. Ashoor, B.B. et al., Principles and applications of direct contact membrane distillation (DCMD): A comprehensive review. *Desalination*, 2016. **398**: pp. 222–246.
32. El-Bourawi, M.S. et al., A framework for better understanding membrane distillation separation process. *Journal of Membrane Science*, 2006. **285**(1–2): pp. 4–29.
33. Vanneste, J. et al., Novel thermal efficiency-based model for determination of thermal conductivity of membrane distillation membranes. *Journal of Membrane Science*, 2018. **548**: pp. 298–308.
34. Cabassud, C. and D. Wirth, Membrane distillation for water desalination: How to chose an appropriate membrane? *Desalination*, 2003. **157**(1): pp. 307–314.
35. Eykens, L. et al., Influence of membrane thickness and process conditions on direct contact membrane distillation at different salinities. *Journal of Membrane Science*, 2016. **498**: pp. 353–364.
36. Gostoli, C., G.C. Sarti, and S. Matulli, Low temperature distillation through hydrophobic membranes. *Separation Science and Technology*, 1987. **22**(2–3): pp. 855–872.
37. Martínez, L. and J.M. Rodríguez-Maroto, Membrane thickness reduction effects on direct contact membrane distillation performance. *Journal of Membrane Science*, 2008. **312**(1–2): pp. 143–156.
38. Wu, H.Y., R. Wang, and R.W. Field, Direct contact membrane distillation: An experimental and analytical investigation of the effect of membrane thickness upon transmembrane flux. *Journal of Membrane Science*, 2014. **470**: pp. 257–265.
39. Ali, M.I. et al., Effects of membrane properties on water production cost in small scale membrane distillation systems. *Desalination*, 2012. **306**: pp. 60–71.

40. Swaminathan, J. et al., Energy efficiency of membrane distillation up to high salinity: Evaluating critical system size and optimal membrane thickness. *Applied Energy*, 2018. **211**: pp. 715–734.
41. González, D., J. Amigo, and F. Suárez, Membrane distillation: Perspectives for sustainable and improved desalination. *Renewable and Sustainable Energy Reviews*, 2017. **80**: pp. 238–259.
42. Kalla, S., S. Upadhyaya, and K. Singh, Principles and advancements of air gap membrane distillation. *Reviews in Chemical Engineering*. 2018, Berlin, Germany: De Gruyter.
43. Gryta, M., Effectiveness of water desalination by membrane distillation process. *Membranes*, 2012. **2**(3): pp. 415–429.
44. Xu, J. et al., Effect of operating parameters and membrane characteristics on air gap membrane distillation performance for the treatment of highly saline water. *Journal of Membrane Science*, 2016. **512**: pp. 73–82.
45. Cath, T.Y., V.D. Adams, and A.E. Childress, Experimental study of desalination using direct contact membrane distillation: A new approach to flux enhancement. *Journal of Membrane Science*, 2004. **228**(1): pp. 5–16.
46. He, K. et al., Production of drinking water from saline water by direct contact membrane distillation (DCMD). *Journal of Industrial and Engineering Chemistry*, 2011. **17**(1): pp. 41–48.
47. Karanikola, V. et al., Sweeping gas membrane distillation: Numerical simulation of mass and heat transfer in a hollow fiber membrane module. *Journal of Membrane Science*, 2015. **483**: pp. 15–24.
48. Charfi, K., M. Khayet, and M.J. Safi, Numerical simulation and experimental studies on heat and mass transfer using sweeping gas membrane distillation. *Desalination*, 2010. **259**(1–3): pp. 84–96.
49. Guillén-Burrieza, E. et al., Experimental evaluation of two pilot-scale membrane distillation modules used for solar desalination. *Journal of Membrane Science*, 2012. *409–410*: pp. 264–275.
50. Swaminathan, J. and J.H. Lienhard, Design and operation of membrane distillation with feed recirculation for high recovery brine concentration. *Desalination*, 2018. **445**: pp. 51–62.
51. Lokare, O.R. et al., Importance of feed recirculation for the overall energy consumption in membrane distillation systems. *Desalination*, 2018. **428**: pp. 250–254.
52. Smolders, K. and A.C.M. Franken, Terminology for membrane distillation. *Desalination*, 1989. **72**(3): pp. 249–262.
53. Qtaishat, M. et al., Heat and mass transfer analysis in direct contact membrane distillation. *Desalination*, 2008. **219**(1–3): pp. 272–292.
54. Zhao, S. et al., Condensation, re-evaporation and associated heat transfer in membrane evaporation and sweeping gas membrane distillation. *Journal of Membrane Science*, 2015. **475**: pp. 445–454.
55. Shrivastava, A., S. Kumar, and E.L. Cussler, Predicting the effect of membrane spacers on mass transfer. *Journal of Membrane Science*, 2008. **323**(2): pp. 247–256.
56. Chernyshov, M.N., G.W. Meindersma, and A.B. de Haan, Comparison of spacers for temperature polarization reduction in air gap membrane distillation. *Desalination*, 2005. **183**(1–3): pp. 363–374.

57. Phattaranawik, J., R. Jiraratananon, and A.G. Fane, Effects of net-type spacers on heat and mass transfer in direct contact membrane distillation and comparison with ultrafiltration studies. *Journal of Membrane Science*, 2003. **217**(1): pp. 193–206.
58. Phattaranawik, J. et al., Mass flux enhancement using spacer filled channels in direct contact membrane distillation. *Journal of Membrane Science*, 2001. **187**(1): pp. 193–201.
59. Da Costa, A.R., A.G. Fane, and D.E. Wiley, Spacer characterization and pressure drop modelling in spacer-filled channels for ultrafiltration. *Journal of Membrane Science*, 1994. **87**(1): pp. 79–98.
60. Yun, Y.B. et al., Effects of channel spacers on direct contact membrane distillation. *Desalination and Water Treatment*, 2011. **34**(1–3): pp. 63–69.
61. Shakaib, M. et al., A CFD study on the effect of spacer orientation on temperature polarization in membrane distillation modules. *Desalination*, 2012. **284**: pp. 332–340.
62. Al-Sharif, S. et al., Modelling flow and heat transfer in spacer-filled membrane distillation channels using open source CFD code. *Desalination*, 2013. **311**: pp. 103–112.
63. Shakaib, M. et al., A CFD study of heat transfer through spacer channels of membrane distillation modules. *Desalination and Water Treatment*, 2013. **51**(16–18): pp. 3662–3674.
64. Tamburini, A. et al., Investigation of heat transfer in spacer-filled channels by experiments and direct numerical simulations. *International Journal of Heat and Mass Transfer*, 2016. **93**: pp. 1190–1205.
65. Chang, H. et al., Simulation study of transfer characteristics for spacer-filled membrane distillation desalination modules. *Applied Energy*, 2017. **185**: pp. 2045–2057.
66. Hagedorn, A. et al., Membrane and spacer evaluation with respect to future module design in membrane distillation. *Desalination*, 2017. **413**: pp. 154–167.
67. La Cerva, M. et al., On some issues in the computational modelling of spacer-filled channels for membrane distillation. *Desalination*, 2017. **411**: pp. 101–111.
68. Katsandri, A., A theoretical analysis of a spacer filled flat plate membrane distillation modules using CFD: Part I: Velocity and shear stress analysis. *Desalination*, 2017. **408**: pp. 145–165.
69. Katsandri, A., A theoretical analysis of a spacer filled flat plate membrane distillation modules using CFD: Part II: Temperature polarisation analysis. *Desalination*, 2017. **408**: pp. 166–180.
70. Taamneh, Y. and K. Bataineh, Improving the performance of direct contact membrane distillation utilizing spacer-filled channel. *Desalination*, 2017. **408**: pp. 25–35.
71. Siddiqui, A. et al., Porosity of spacer-filled channels in spiral-wound membrane systems: Quantification methods and impact on hydraulic characterization. *Water Research*, 2017. **119**: pp. 304–311.
72. Seo, J., Y.M. Kim, and J.H. Kim, Spacer optimization strategy for direct contact membrane distillation: Shapes, configurations, diameters, and numbers of spacer filaments. *Desalination*, 2017. **417**: pp. 9–18.
73. Ponzio, F.N. et al., Experimental and computational investigation of heat transfer in channels filled by woven spacers. *International Journal of Heat and Mass Transfer*, 2017. **104**: pp. 163–177.

74. Albeirutty, M. et al., An experimental study for the characterization of fluid dynamics and heat transport within the spacer-filled channels of membrane distillation modules. *Desalination*, 2018. **430**: pp. 136–146.
75. Thomas, N. et al., 3D printed triply periodic minimal surfaces as spacers for enhanced heat and mass transfer in membrane distillation. *Desalination*, 2018. **443**: pp. 256–271.
76. Kim, Y.-D. et al., Effect of non-woven net spacer on a direct contact membrane distillation performance: Experimental and theoretical studies. *Journal of Membrane Science*, 2018. **564**: pp. 193–203.
77. Ding, Z., R. Ma, and A.G. Fane, A new model for mass transfer in direct contact membrane distillation. *Desalination*, 2003. **151**(3): pp. 217–227.
78. Ho, C.D. et al., Performance improvement on distillate flux of countercurrent-flow direct contact membrane distillation systems. *Desalination*, 2014. **338**: pp. 26–32.
79. Alsaadi, A.S. et al., Modeling of air-gap membrane distillation process: A theoretical and experimental study. *Journal of Membrane Science*, 2013. **445**: pp. 53–65.
80. Lienhard V, J.H. *Thermophysical properties of seawater – http://web.mit.edu/seawater/*. 2017 [cited October 13, 2018]; Available from: http://web.mit.edu/seawater/.
81. Martinez, L. and J.M. Rodriguez-Maroto, On transport resistances in direct contact membrane distillation. *Journal of Membrane Science*, 2007. **295**(1–2): pp. 28–39.
82. Zhongwei, D., L. Liying, and M. Runyu, Study on the effect of flow maldistribution on the performance of the hollow fiber modules used in membrane distillation. *Journal of Membrane Science*, 2003. **215**(1): pp. 11–23.
83. Gabelman, A. and S.T. Hwang, Hollow fiber membrane contactors. *Journal of Membrane Science*, 1999. **159**(1–2): pp. 61–106.
84. Francis, L. et al., Performance of different hollow fiber membranes for seawater desalination using membrane distillation. *Desalination and Water Treatment*, 2014. **55**(10): pp. 2786–2791.
85. Chung, S. et al., Evaluation method of membrane performance in membrane distillation process for seawater desalination. *Environmental Technology*, 2014. **35**(17): pp. 2147–2152.
86. Yu, H. et al., Numerical simulation of heat and mass transfer in direct membrane distillation in a hollow fiber module with laminar flow. *Journal of Membrane Science*, 2011. **384**(1–2): pp. 107–116.
87. Wang, Z.S. et al., Analysis of DCMD-based hollow fiber membrane heat exchanger. *Journal of Chemical Engineering of Japan*, 2013. **46**(9): pp. 573–582.
88. Yang, X. et al., Performance improvement of PVDF hollow fiber-based membrane distillation process. *Journal of Membrane Science*, 2011. **369**(1–2): pp. 437–447.
89. Yazgan-Birgi, P., M.I. Hassan Ali, and H.A. Arafat, Comparative performance assessment of flat sheet and hollow fiber DCMD processes using CFD modeling. *Separation and Purification Technology*, 2019. **212**: pp. 709–722.
90. Al-Khatib, A., *An Experimental Comparison of Performance between Flat Sheet and Hollow Fiber Membrane Modules in Direct Contact Membrane Distillation System for Desalination of Seawater*. 2016, Ann Arbor, MI: Qatar University (Qatar). p. 238.
91. Francis, L. et al., Performance evaluation of the DCMD desalination process under bench scale and large scale module operating conditions. *Journal of Membrane Science*, 2014. **455**: pp. 103–112.

92. Nghiem, L.D. et al., Treatment of saline aqueous solutions using direct contact membrane distillation. *Desalination and Water Treatment*, 2011. **32**(1–3): pp. 234–241.
93. Zhang, J. et al., Identification of material and physical features of membrane distillation membranes for high performance desalination. *Journal of Membrane Science*, 2010. **349**(1–2): pp. 295–303.
94. Adham, S. et al., Application of membrane distillation for desalting brines from thermal desalination plants. *Desalination*, 2013. **314**: pp. 101–108.
95. Eleiwi, F. et al., Dynamic modeling and experimental validation for direct contact membrane distillation (DCMD) process. *Desalination*, 2016. **384**: pp. 1–11.
96. Shirazi, M.M.A., A. Kargari, and M. Tabatabaei, Evaluation of commercial PTFE membranes in desalination by direct contact membrane distillation. *Chemical Engineering and Processing: Process Intensification*, 2014. **76**: pp. 16–25.
97. Shim, W.G. et al., Solar energy assisted direct contact membrane distillation (DCMD) process for seawater desalination. *Separation and Purification Technology*, 2015. **143**: pp. 94–104.
98. Qu, D. et al., Comparison of hollow fiber module designs in membrane distillation process employed lumen-side and shell-side feed. *Desalination and Water Treatment*, 2016. **57**(17): pp. 7700–7710.
99. Singh, D. and K.K. Sirkar, Performance of PVDF flat membranes and hollow fibers in desalination by direct contact membrane distillation at high temperatures. *Separation and Purification Technology*, 2017. **187**: pp. 264–273.
100. Yang, X., R. Wang, and A.G. Fane, Novel designs for improving the performance of hollow fiber membrane distillation modules. *Journal of Membrane Science*, 2011. **384**(1–2): pp. 52–62.
101. Yang, X. et al., Analysis of the effect of turbulence promoters in hollow fiber membrane distillation modules by computational fluid dynamic (CFD) simulations. *Journal of Membrane Science*, 2012. *415–416*: pp. 758–769.
102. Chung, H.W. et al., Multistage vacuum membrane distillation (MSVMD) systems for high salinity applications. *Journal of Membrane Science*, 2016. **497**: pp. 128–141.
103. Alsaadi, A.S. et al., Flashed-feed VMD configuration as a novel method for eliminating temperature polarization effect and enhancing water vapor flux. *Journal of Membrane Science*, 2018. **563**: pp. 175–182.
104. Francis, L. et al., Material gap membrane distillation: A new design for water vapor flux enhancement. *Journal of Membrane Science*, 2013. **448**: pp. 240–247.
105. He, F., J. Gilron, and K.K. Sirkar, High water recovery in direct contact membrane distillation using a series of cascades. *Desalination*, 2013. **323**: pp. 48–54.
106. Summers, E.K., H.A. Arafat, and J.H. Lienhard, Energy efficiency comparison of single-stage membrane distillation (MD) desalination cycles in different configurations. *Desalination*, 2012. **290**: pp. 54–66.
107. Koschikowski, J., M. Wieghaus, and M. Rommel, Solar thermal-driven desalination plants based on membrane distillation. *Desalination*, 2003. **156**(1): pp. 295–304.
108. Gilron, J., L. Song, and K.K. Sirkar, Design for cascade of crossflow direct contact membrane distillation. *Industrial & Engineering Chemistry Research*, 2007. **46**(8): pp. 2324–2334.

109. Zuo, G. et al., Energy efficiency evaluation and economic analyses of direct contact membrane distillation system using Aspen Plus. *Desalination*, 2011. **283**: pp. 237–244.
110. Swaminathan, J. et al., Membrane distillation model based on heat exchanger theory and configuration comparison. *Applied Energy*, 2016. **184**: pp. 491–505.
111. Zhang, Y. et al., Review of thermal efficiency and heat recycling in membrane distillation processes. *Desalination*, 2015. **367**: pp. 223–239.
112. Deshmukh, A. and M. Elimelech, Understanding the impact of membrane properties and transport phenomena on the energetic performance of membrane distillation desalination. *Journal of Membrane Science*, 2017. **539**: pp. 458–474.
113. Elimelech, M. and W.A. Phillip, The future of seawater desalination: Energy, technology, and the environment. *Science*, 2011. **333**(6043): pp. 712–717.
114. Criscuoli, A., M.C. Carnevale, and E. Drioli, Evaluation of energy requirements in membrane distillation. *Chemical Engineering and Processing*, 2008. **47**(7): pp. 1098–1105.
115. Khayet, M., Solar desalination by membrane distillation: Dispersion in energy consumption analysis and water production costs (a review). *Desalination*, 2013. **308**: pp. 89–101.
116. Al-Obaidani, S. et al., Potential of membrane distillation in seawater desalination: Thermal efficiency, sensitivity study and cost estimation. *Journal of Membrane Science*, 2008. **323**(1): pp. 85–98.
117. Alklaibi, A.M. and N. Lior, Comparative study of direct-contact and air-gap membrane distillation processes. *Industrial & Engineering Chemistry Research*, 2007. **46**(2): pp. 584–590.
118. Lawson, K.W. and D.R. Lloyd, Membrane distillation. 2. Direct contact MD. *Journal of Membrane Science*, 1996. **120**(1): pp. 123–133.
119. Lienhard, J.H.M., Karan H., Sharqawy, M. H.; Thiel, G. P., Thermodynamics, exergy, and energy efficiency in desalination systems, eds. H.A. Arafat, *Desalination Sustainability: A Technical, Socioeconomic, and Environmental Approach*, . 2017, Cambridge, MA: Elsevier.
120. Karakulski, K., M. Gryta, and A. Morawski, Membrane processes used for potable water quality improvement. *Desalination*, 2002. **145**(1): pp. 315–319.
121. Schneider, K. et al., Membranes and modules for transmembrane distillation. *Journal of Membrane Science*, 1988. **39**(1): pp. 25–42.
122. Rodriguez, J.R.B. et al., Distilled and drinkable water quality produced by solar membrane distillation technology. *Desalination and Water Treatment*, 2013. **51**(4–6): pp. 1265–1271.
123. Andersson, S.I., N. Kjellander, and B. Rodesjö, Design and field tests of a new membrane distillation desalination process. *Desalination*, 1985. **56**: pp. 345–354.
124. Song, L. et al., Pilot plant studies of novel membranes and devices for direct contact membrane distillation-based desalination. *Journal of Membrane Science*, 2008. **323**(2): pp. 257–270.
125. Meindersma, G.W., C.M. Guijt, and A.B. de Haan, Desalination and water recycling by air gap membrane distillation. *Desalination*, 2006. **187**(1–3): pp. 291–301.

126. Minier-Matar, J. et al., Field evaluation of membrane distillation technologies for desalination of highly saline brines. *Desalination*, 2014. **351**: pp. 101–108.
127. Gil, J.D. et al., A feedback control system with reference governor for a solar membrane distillation pilot facility. *Renewable Energy*, 2018. **120**: pp. 536–549.
128. Cassard, H.M. and H.G. Park, How to select the optimal membrane distillation system for industrial applications. *Journal of Membrane Science*, 2018. **565**: pp. 402–410.

6

Membrane Fouling and Scaling in Membrane Distillation

6.1 Introduction

Membranes in separation processes can be prone to fouling leading to performance deterioration. In membrane distillation for desalination, fouling does occur and this is manifested experimentally in the form of permeate flux decline with a potential deterioration of the quality of distillate. This can be due to membrane "wetting" where dissolved salt ions are carried through the membrane pores to the distillate side of the membrane, thus increasing the total dissolved solids (TDS) content and reducing the salt rejection. In this chapter we will shed some light on the factors that lead to performance deterioration in membrane distillation due to fouling. The topic of membrane fouling in desalination is quite complex and has been investigated extensively in reverse osmosis [1–7]. It has also attracted attention in membrane distillation for desalination in recent years due to the resurge in interest in this technology for the production of freshwater [8–11]. While membrane technologies for desalination in reverse osmosis (RO) [7,12,13] and membrane distillation (MD) [14–17] are different, it is instructive to learn lessons gained from the long and extensive experience in reverse osmosis as there are similarities in concepts such as "concentration polarization" near the membrane surface on the feed side. It is important to bear in mind that while RO has been commercially exploited for decades with a tremendous practical experience gained, MD is yet to be deployed on a commercial scale, with much of the work done in small-scale laboratory MD modules. Hence reference to some relevant lessons learned in RO will be invoked in this chapter where appropriate. It is, however, instructive to note that the membrane fouling and wetting mechanisms are still speculative but increasingly being supported by experimental evidence and membrane autopsy post use [8,10,18–22].

The most frequently reported speculative mechanism for membrane fouling refers to material depositing on the membrane surface, known as surface fouling, or material depositing inside membrane pores, known as pore blocking. Figure 6.1 depicts a diagram of these fouling mechanisms. In either case,

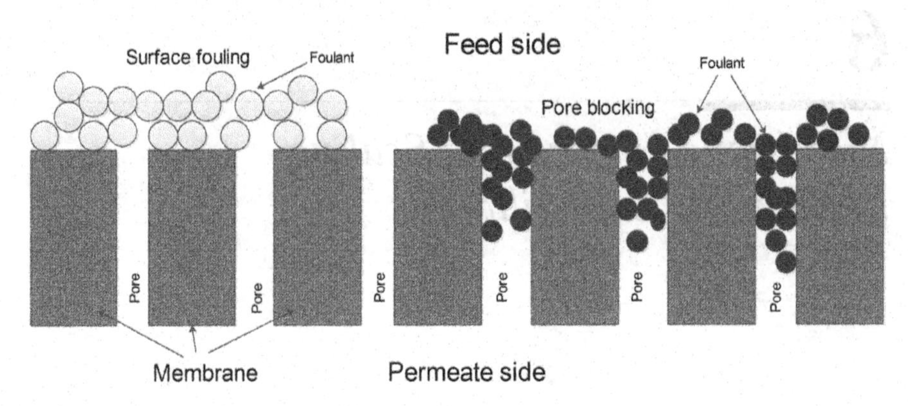

FIGURE 6.1
Representation of fouling mechanisms in membrane distillation.

a permeate flux decline is observed due to a reduction in "active" membrane area for vapor transport from the feed side to the permeate side. In practice, it is possible that both mechanisms occur and there is simply no experimental evidence reported and supported by membrane autopsy because it is currently nearly impossible to gain access to the membrane surface during the runs, and no work has been reported on periodic membrane inspections involving extracting the used membrane and re-using it in further runs. This situation demonstrates how difficult it is to ascertain fouling mechanisms in MD. In addition, the literature also reports membrane wetting when the permeate flux declines, indicating some sort of changes in the properties of the MD membrane surface that reduce or even destroy its hydrophobicity. The wetting phenomenon is often detected when the distillate quality deteriorates. There have been attempts to speculate on the potential mechanisms for membrane wetting in MD. Rezaei et al. [18] published a recent review on membrane wetting in MD and highlighted possible mechanisms with accompanying flux trends:

- Transmembrane pressure exceeding the membrane LEP. This can be caused by excessive operating pressure and presence of very large membrane pores.
- Capillary condensation. Often due to loss of temperature gradient, possibly due to temporary shutdowns or variable operating temperatures that can reduce the saturation pressure of the permeate vapor leading to condensation. It has to be recalled that the terminology for MD specifically excludes capillary condensation [23].
- Scale deposition (inorganic fouling). Mineral deposits on the membrane or pore surface that lead to reduction or destruction of hydrophobicity.

- Organic fouling. Deposits of organic matter that can form attractive forces between hydrophobic materials within the aqueous system or promote a reduction in surface tension.
- Surfactants. The presence of these in the feed will lower the surface tension and promote wetting. This is well known in MD when membranes are deliberately wetted with isopropyl alcohol and water solutions to recondition membrane pores, rinse them and dry them for re-use. The literature also has studies on wetting due to alcohol solutions that reduce the liquid entry pressure [24].
- Membrane degradation. This can be due to chemical changes to the membrane that leads to formation of hydrophilic functional groups or can be due to mechanical degradation that can form cracks and larger pores.

When membrane wetting occurs and is observed through an increase in permeate conductivity, the permeate can decrease if only surface wetting occurs or actually increase when partial or full pore wetting occurs [18].

Factors that promote membrane fouling can be numerous, but the literature emphasizes four of these:

- The nature of the feed
- The foulant characteristics
- The membrane properties
- The operating conditions

Figure 6.2 depicts a diagram showing a partially fouled membrane and possible main factors that can lead to fouling. The SEM image shown in Figure 6.2 shows that the membrane lost part of its active surface through material deposition, and that obviously will lead to a reduced permeate flux. However, SEM imaging alone cannot provide information about the possibility of membrane wetting. Additional membrane surface analysis techniques such as atomic force microscopy (AFM) can provide data on the membrane surface roughness (before and after use), which is closely linked to hydrophobicity and can be used in association with membrane contact angle measurements [22,25,26] to boost the level of confidence in tracking hydrophobicity loss.

In desalination using MD, the nature of foulants reported, i.e. whether fouling is caused by inorganic matter or by organic matter, including biomass, indicates a domination of inorganic matter, and this fouling is often referred to as "scaling." However, a limited number of papers also reported on "biofouling" in MD, but caution must exercised and attention be given to the nature of the feed, as indicated in Figure 6.2. This point is crucial since fouling studies reported in the literature include a great variety of feeds, ranging from tap water to seawater, passing through lake water

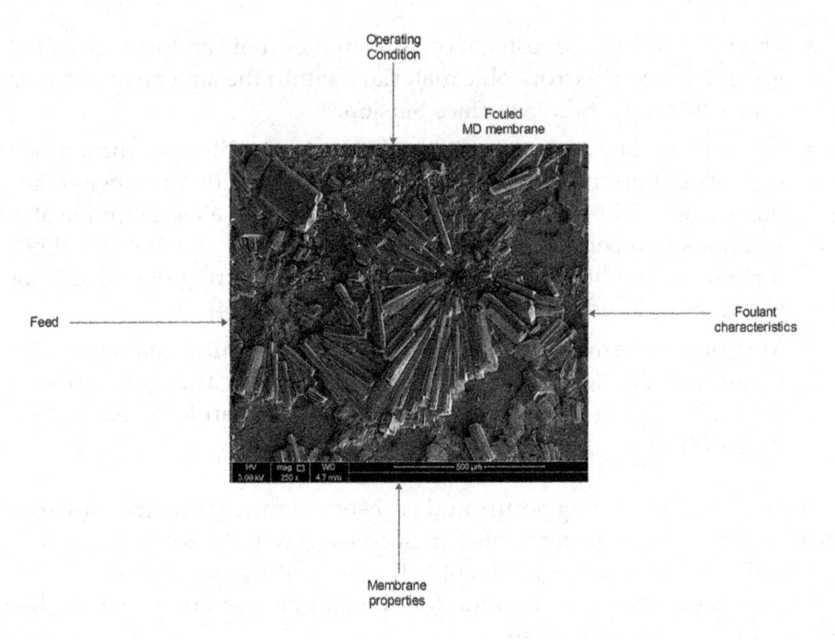

FIGURE 6.2
Diagram showing a partially fouled membrane and possible factors that can contribute to fouling in membrane distillation in desalination.

and wastewater. These feeds have substantially different constituents and also have contents that can inhibit biomass growth (like high salinity). Furthermore, MD is essentially a thermal process where feed temperatures can range from 40°C to 80°C. At the higher end of the temperature range, perhaps above 50°C, biomass growth will be marginal or eliminated as cells are simply destroyed [27]. However, for MD feed from a lake, a recent publication reported biomass growth at temperatures between 40°C and 60°C [19]. A recent thesis also excluded the possibility of significant biofouling in membrane fouling in MD with high salinity seawater feed [21]. In a MD project closely related to [21] some SEM evidence of destroyed plankton was observed at MD runs between 50°C and 70°C, as shown in Figure 6.3. There was no evidence of cell growth on membranes used (neither bacterial nor plankton) with high salinity Gulf seawater feed (40,000–42,000 ppm).

Hence for seawater desalination, and at temperatures above 40°C–45°C, the major foulants appear to be inorganic and the fouling can be called scaling, assuming that the feed is free from organic pollutants. This point is important. That is normally the case in sampling of seawater for experimental MD work. The sampling place in seawater must be assured to be free from any organic pollution.

For MD applicants other than desalination, like water treatment, Gryta [28] reported some membrane fouling results from a variety of feeds: bilge water

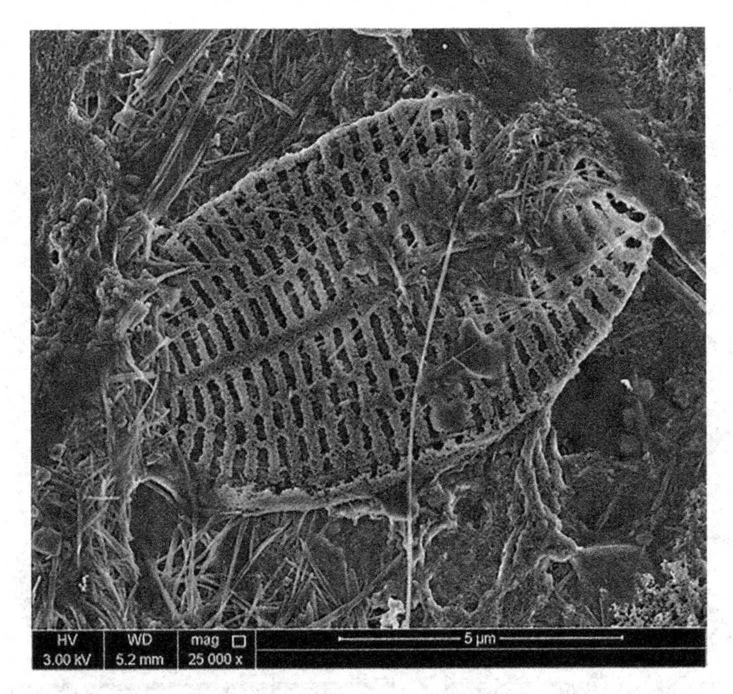

FIGURE 6.3
High resolution SEM of a used membrane showing destroyed sea plankton.

(oily content), saline wastewater from meat processing (where NaCl is added to meats and fish) and tap water (hard water with traces of organic matter).

For the case of purely inorganic scaling in MD, what would be the nature of the foulant? The answer obviously would be the nature of the feed. The literature reported membrane scaling for tap water or seawater, passing through synthetic solutions of various salinities and salt contents as well as saline industrial waters [8,10,21,22,29–36]. Some of the most widely used tools used to identify inorganic foulants (scaling matter) include SEM imaging (crystal shapes) in association with EDS (Energy-dispersive X-ray spectroscopy, also reported as EDX). Figures 6.4 and 6.5 depict a MD membrane fouled (scaled) from high salinity seawater feed from the Gulf and showing well known crystals of calcium carbonate and calcium sulfate in Figure 6.4 and their EDS fingerprint in Figure 6.5. Figure 6.5 also shows some magnesium present is normal for seawater.

A topic often overlooked or unreported in fouling/scaling in MD desalination is the concept of "soft" or "hard" fouling. Soft fouling refers to material that does not adhere permanently to the membrane surface and can be washed away by rinsing the membrane online between runs to restore a declining permeate flux, while hard fouling refers to material that adheres to the membrane surface and cannot be washed away by simple rinsing online. The concept of soft scaling is perhaps overlooked in long term runs in the

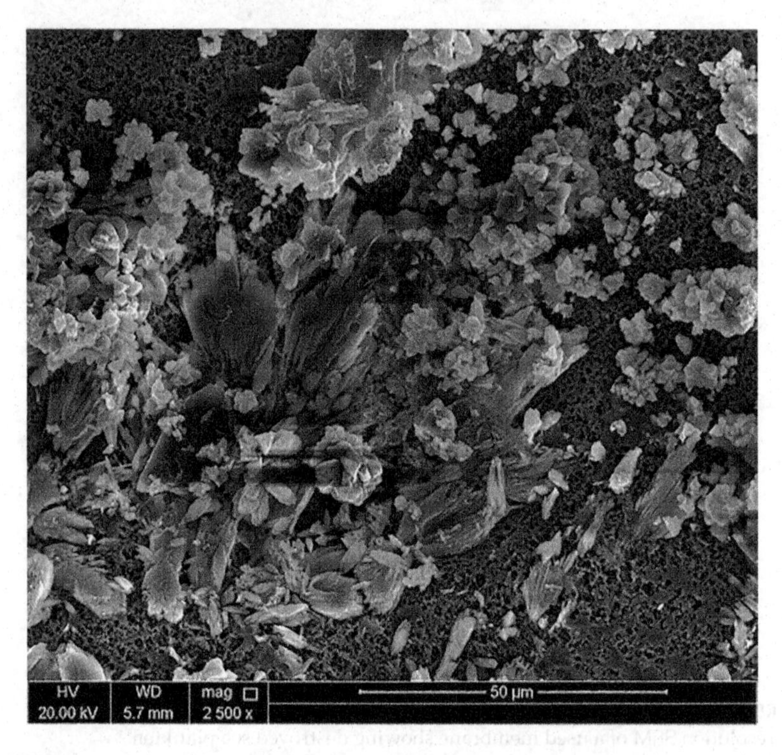

FIGURE 6.4
SEM of a MD membrane fouled (scaled) from high salinity seawater feed from the Arabian Gulf. The crystals of calcium carbonate and calcium sulfate are clearly visible.

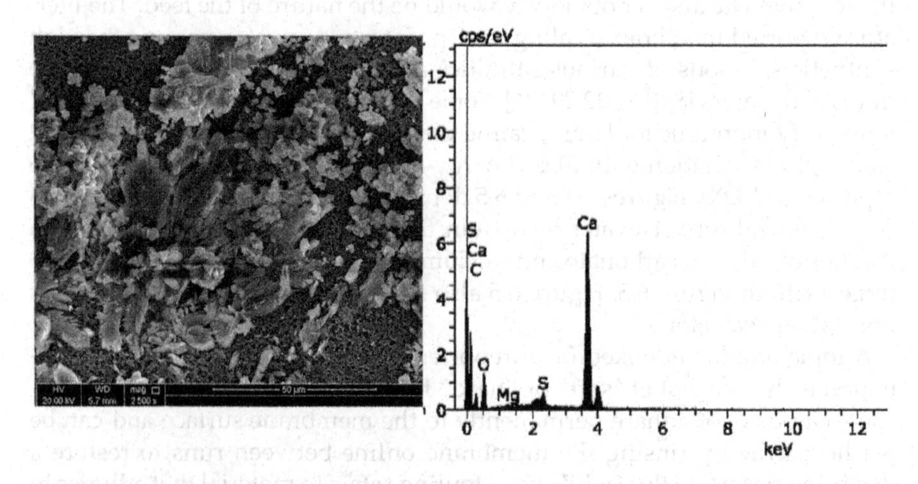

FIGURE 6.5
SEM and EDS of a used and fouled MD membrane. The fingerprint of calcium carbonate and calcium sulfate is clearly seen.

laboratory as the material can easily detach itself and flow away from the membrane. The only evidence of this could be slight fluctuations of permeate flux (down then up), in contrast with hard scaling that always leads to continuously declining flux. Soft scaling can be easily detected when operating the MD module intermittently (batch mode) under identical conditions and recording the permeate flux. Gryta reported some evidence of this phenomenon [28]. Al-Obaidly also observed flux decline fluctuations in a batch mode study on membrane fouling [22].

Let us focus on some aspects of flux decline and membrane fouling.

6.2 Flux and Flux Decline in Membrane Distillation

In MD, the permeate flux is described by equation (4.1), as stated in Chapter 4:

$$J = C_m \left(P_{fm} - P_{pm} \right) \tag{4.1}$$

where J is the permeate flux, C_m is the membrane coefficient that is reported independent of process conditions in most cases [37], and P_{fm} and P_{pm} and the vapor pressure values on the membrane surface at feed and permeate side respectively. While the vapor pressure is traditionally estimated from Antoine equation based correlations for pure water or very dilute solutions, it can be better estimated under wide ranges of conditions using the latest available correlations for seawater feed from the Massachusetts Institute of Technology (MIT) group led by Lienhard [38]. These correlations are updates of previously published ones by the MIT group [39]. The membrane coefficient C_m, despite being reported as nearly constant for specific membranes, is really a lumped parameter encapsulating the membrane morphological parameters such as porosity and tortuosity as well as membrane thickness. Because the water vapor transport (flux) across the membrane distillation (MD) microporous membrane matrix is best described by the so-called dusty gas model, the membrane coefficient has been reported to be some function of the membrane properties [40,41]. The form of the dusty gas model that applies depends on the MD system configuration where the permeate side impacts the way water vapor is transported through the porous membrane [40,41]. Consequently, the membrane coefficient (sometimes called membrane permeability), C_m, can be estimated from the methods described by Khayet et al. [42] or determined experimentally through measured permeate flux values and equation (4.1). When membrane fouling occurs, automatically the permeate flux declines. The extent of flux reduction will obviously be related to the severity of fouling. It is worth noting at this stage that fouling leads to material deposit on the membrane surface with a possibility to also

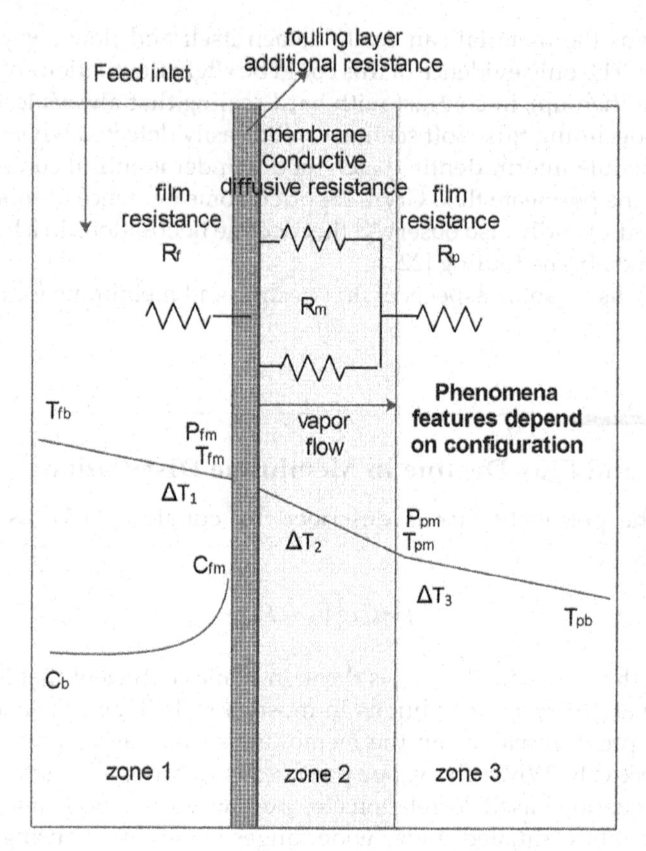

FIGURE 6.6
Additional transport resistance due to fouling layer on the membrane surface.

deposit inside the membrane pores (hot side), as we have seen in Figure 6.1. Let us refer to Figure 5.1 to see the effect of having material deposit on the membrane surface (hot side). It is clear that we need to account for the additional resistance that the deposited material cause and this is now depicted in Figure 6.6. The effect of deposited material on the membrane surface in the hot side can lead to:

- Reduction of the membrane active area (pore blocking).
- Additional transport resistance the permeate vapor has to go through. This depends on the fouling layer porosity. If the deposit porosity is low or zero, the flux will simply drop to zero.
- Increased thermal conductivity resistance immediately after the film resistance (thermal boundary layer), with a net effect of enhancing the temperature polarization further.

The combination of the above points will be the causes of flux reduction. This is the conceptual description of fouling in membrane distillation. Let us explore a number of literature reports on flux decline due to fouling.

He et al. reported work on fouling in DCMD using real seawater [31] and a PTFE membrane with 0.22 μ pore size. They observed a flux reduction from 23.76 to 14.36 L/m²·h (roughly 40% decline in flux) over a period of 30 days.

Al-Obaidli [22] reported in a thesis work on fouling in DCMD using three types of membranes with real Gulf seawater (TDS of around 41200 ppm), namely PTFE (0.22 μ pores) and polypropylene (PP, 0.22 and 0.45 μ pores) and operating the MD system at 75/20°C for the feed and permeate temperatures respectively. After initial permeate fluxes of 50.5, 50.3 and 38.3 L/m²·h for PP (0.22 μm) membrane, PP (0.45 μm) membrane and PTFE (0.22 μm) membrane, reductions in flux of 60%, 97% and 94% for PP membrane of 0.22 μm, PP membrane of 0.45 μm and PTFE membrane of 0.22 μm, respectively, after 19, 30 and 25 h of operation, respectively.

Younis [43] reported in a thesis work on DCMD using real Gulf seawater and PP membrane (0.2 μ) that permeate flux declined by 74% and 92% for feed temperatures of 60°C and 70°C respectively over a period of 91 h.

6.3 Fouling and Scaling in Membrane Distillation

How does fouling occur during membrane distillation of hot seawater feed? We have some preliminary clues pointing at the possibility that concentration polarization on the membrane surface may well be a necessary first step. The equation for the concentration polarization coefficient, CPC is given by:

$$CPC = \frac{S_{fm}}{S_f} = \exp\left(\frac{J}{\rho_f K}\right) \tag{6.1}$$

where S_{fm} and S_f are the salt concentrations on the membrane surface and in bulk solution respectively, J is the permeate flux, ρ_f is the feed density and K the mass transfer coefficient of salt. Equation (6.1) indicates that an increase in flux will also increase the CPC, making the salt concentration on the membrane surface higher. How much higher? That very much depends on the operating conditions, the MD module design and the membrane itself. It is important to bear in mind that the CPC can only be estimated through modeling. Unfortunately, the literature lacks hard data on the values of S_{fm}. The literature has consistently shown that the scaling material on the membrane surface for seawater feed is a mixture of salts dominated by salts of calcium (carbonates and sulfates) [10,21,22,29,31]. Gryta [44,45] also showed that carbonates and sulfates of calcium do precipitate on membrane surfaces at certain conditions. For salts of calcium and any other minor cations in the feed to precipitate on the membrane surface, there must be a phenomenon where salt saturation occurs and crystallization taking place, as shown in Figure 6.4. The literature lacks experimental evidence of salt concentration saturation on the membrane surface followed by onset of crystallization because if these occur, they do so in a very thin layer above the membrane and it would be extremely difficult to observe. One can only speculate on the scaling mechanism at this stage and there is an urgent need to elucidate this phenomenon. Additional clues that support the possibility of this mechanism occurring comes from experiments reported in the literature where anti-scalant was added to the feed seawater and successfully blocking flux decline [21,30,34]. It is interesting to note that a great deal of studies have shown the nature of scale formation in MD [8,20,21,29,36,46].

6.4 Membrane Autopsy Techniques in Membrane Distillation

Membrane autopsy has been pioneered in reverse osmosis and a valuable experience gained [47–52]. It is also gaining importance in MD studies related to membrane fouling [53].

Recent work on membrane fouling in MD reported a novel technique referred to as "electrically conductive membranes for in situ fouling detection in membrane distillation using impedance spectroscopy" [54]. Due to the duration of experiments on membrane fouling in MD and the cost of comprehensive membrane autopsy, very little work has been published. Much of the fouling and used membrane characterization work refer to limited techniques like contact angle, SEM-EDX and occasionally atomic force spectroscopy.

6.5 Membrane Wetting and Distillate Quality Deterioration

We recall from Chapter 3 that Young's contact angle and can be found in Young's equation [55]:

$$\gamma_{lv} \cos \theta = \gamma_{sv} - \gamma_{sl} \tag{3.1}$$

where γ_{lv}, γ_{sv} and γ_{sl} represent the liquid vapor, solid vapor and solid liquid interfacial surface tensions respectively and that the liquid entry pressure (LEP) is defined by equation (3.2) [56–58]:

$$\text{LEP} = \frac{-2B\gamma_l}{d_{max}} \cos(\theta) \tag{3.2}$$

where B is a geometric factor, γ_l is the surface tension of the liquid solution, θ is the contact angle and d_{max} is the largest pore size of the membrane.

When the membrane flux performance declines and starts showing signs of wetting (distillate quality deteriorates), the membrane contact angle θ (in radians) drops, and from equation (3.2), the liquid entry pressure (LEP) for that membrane also drops, potentially leading to a "positive feedback" effect, thus accelerating the membrane wetting process. An experimental evidence of rapid wetting following severe fouling in polypropylene (PP) and PTFE membranes was reported [22]. The larger pores PP membrane showed the most fouling and wetting. More work is needed to verify the extent of "positive feedback" in membrane wetting. The following references cover aspects of membrane wetting through permeate quality, contact angle [18,22,24,43,59–64] and the use of a probe to detect wetting [65].

6.6 Fouling Mitigation Measure in Membrane Distillation

Since MD is not deployed commercially yet, there is no large-scale MD facility experience on tackling fouling in MD. On the other hand, some work was done on small laboratory scale and pilot plant scale fouling mitigation. Fouling mitigation can be initiated in a number of ways: Using anti-scalant additives [30,34,35,43,66], treating fouled membranes with acid/chemicals [67] or using fouling resistant membranes. Other forms of mitigating scale formation have been reported in the literature: scaling control with air bubbling in MD [68], air backwash [20] and Ultrasonic mitigation of MD [69]. There are other ways to minimize membrane scaling: tuning operational parameters of the MD unit to minimize concentration polarization on the membrane surface. However, this is of limited use if large scale deployment is an aim.

6.7 Future Directions in Membrane Fouling Resistance Efforts

The ideal objective in membrane distillation would be to develop a system where fouling does not occur significantly. This is not an easy objective to achieve because concentration polarization is inherent to the MD operation and that higher feed temperatures are desirables because they give higher permeate fluxes. Hence the only realistic objective in this matter would be to develop fouling resisting membranes. Some efforts are being deployed in recent years, and this topic will be covered in the next chapter.

6.8 Concluding Remarks

Membrane fouling is a reality to any membrane separation process. In MD, however, this undesirable phenomenon is compounded by the fact that higher permeate fluxes, which is desirable in terms of MD performance, will always create situations where membrane fouling will have a higher potential to occur. Our current knowledge of the mechanism and reasons why fouling material attaches itself to the hydrophobic membrane surface is limited. A greater understanding of these phenomena would undoubtedly lead to more effective and appropriate solutions. For now, we resort to the dream of inventing "superhydrophobic" membranes that are also fouling resistant. That is indeed an ambitious objective.

References

1. Jiang, S., Y. Li, and B.P. Ladewig, A review of reverse osmosis membrane fouling and control strategies.*Science of the Total Environment*, 2017. **595**: pp. 567–583.
2. Khan, M.T. et al., How different is the composition of the fouling layer of wastewater reuse and seawater desalination RO membranes? *Water Research*, 2014. **59**: pp. 271–282.
3. Sim, L.N. et al., A review of fouling indices and monitoring techniques for reverse osmosis. *Desalination*, 2018. **434**: pp. 169–188.
4. Goosen, M.F.A. et al., Fouling of reverse osmosis and ultrafiltration membranes: A critical review. *Separation Science and Technology*, 2005. **39**(10): pp. 2261–2297.
5. Potts, D.E., R.C. Ahlert, and S.S. Wang, A critical review of fouling of reverse osmosis membranes. *Desalination*, 1981. **36**(3): pp. 235–264.
6. Bucs, S.S. et al., Review on strategies for biofouling mitigation in spiral wound membrane systems. *Desalination*, 2018. **434**: pp. 189–197.
7. Lee, K.P., T.C. Arnot, and D. Mattia, A review of reverse osmosis membrane materials for desalination—Development to date and future potential. *Journal of Membrane Science*, 2011. **370**(1–2): pp. 1–22.
8. Naidu, G. et al., A review on fouling of membrane distillation. *Desalination and Water Treatment*, 2016. **57**(22): pp. 10052–10076.
9. Bogler, A., S. Lin, and E. Bar-Zeev, Biofouling of membrane distillation, forward osmosis and pressure retarded osmosis: Principles, impacts and future directions. *Journal of Membrane Science*, 2017. **542**: pp. 378–398.
10. Tijing, L.D. et al., Fouling and its control in membrane distillation: A review. *Journal of Membrane Science*, 2015. **475**: pp. 215–244.
11. Nguyen, Q.-M. and S. Lee, Fouling analysis and control in a DCMD process for SWRO brine. *Desalination*, 2015. **367**: pp. 21–27.
12. Misdan, N., W.J. Lau, and A.F. Ismail, Seawater Reverse Osmosis (SWRO) desalination by thin-film composite membrane—Current development, challenges and future prospects. *Desalination*, 2012. **287**: pp. 228–237.

13. Yang, Z., X.-H. Ma, and C.Y. Tang, Recent development of novel membranes for desalination. *Desalination*, 2018. **434**: pp. 37–59.
14. García-Fernández, L., M. Khayet, and M.C. García-Payo, 11 - Membranes used in membrane distillation: Preparation and characterization, in *Pervaporation, Vapour Permeation and Membrane Distillation*, A. Basile, A. Figoli, and M. Khayet, Editors. 2015, Woodhead Publishing: Oxford. pp. 317–359.
15. Khayet, M. et al., Design of novel direct contact membrane distillation membranes. *Desalination*, 2006. **192**(1–3): pp. 105–111.
16. Makanjuola, O., I. Janajreh, and R. Hashaikeh, Novel technique for fabrication of electrospun membranes with high hydrophobicity retention. *Desalination*, 2018. **436**: pp. 98–106.
17. Duong, H.C. et al., A novel electrospun, hydrophobic, and elastomeric styrene-butadiene-styrene membrane for membrane distillation applications. *Journal of Membrane Science*, 2018. **549**: pp. 420–427.
18. Rezaei, M. et al., Wetting phenomena in membrane distillation: Mechanisms, reversal, and prevention. *Water Research*, 2018. **139**: pp. 329–352.
19. Liu, C. et al., The effect of feed temperature on biofouling development on the MD membrane and its relationship with membrane performance: An especial attention to the microbial community succession. *Journal of Membrane Science*, 2019. **573**: pp. 377–392.
20. Zou, T. et al., Fouling behavior and scaling mitigation strategy of CaSO4 in submerged vacuum membrane distillation. *Desalination*, 2018. **425**: pp. 86–93.
21. Younis, R.F., An investigation into fouling, wetting and their mitigation using commercial Antiscalant in membrane distillation desalination, in Thesis, Master of Science (Environmental Engineering). http://hdl.handle.net/10576/5755 (accessed 09/10/2018). 2017, Qatar University.
22. Al-Obaidli, M.A.A.M., *An Investigation into Hydrophobic Membrane Fouling in Desalination Using Membrane Distillation Technology*. 2015, Qatar University: Qatar University. p. 347.
23. Smolders, K. and A.C.M. Franken, Terminology for membrane distillation. *Desalination*, 1989. **72**(3): pp. 249–262.
24. García-Payo, M.C., M.A. Izquierdo-Gil, and C. Fernández-Pineda, Wetting study of hydrophobic membranes via liquid entry pressure measurements with aqueous alcohol solutions. *Journal of Colloid and Interface Science*, 2000. **230**(2): pp. 420–431.
25. Kim, B.-S. and P. Harriott, Critical entry pressure for liquids in hydrophobic membranes. *Journal of Colloid and Interface Science*, 1987. **115**(1): pp. 1–8.
26. Mittal, K.L. et al. *Contact Angle, Wettability and Adhesion*. Utrecht, the Netherlands: VSP.
27. Krivorot, M. et al., Factors affecting biofilm formation and biofouling in membrane distillation of seawater. *Journal of Membrane Science*, 2011. **376**(1–2): pp. 15–24.
28. Gryta, M., Fouling in direct contact membrane distillation process. *Journal of Membrane Science*, 2008. **325**(1): pp. 383–394.
29. Warsinger, D.M. et al., Scaling and fouling in membrane distillation for desalination applications: A review. *Desalination*, 2015. **356**: pp. 294–313.
30. Peng, Y.L. et al., Effects of anti-scaling and cleaning chemicals on membrane scale in direct contact membrane distillation process for RO brine concentrate. *Separation and Purification Technology*, 2015. **154**: pp. 22–26.
31. He, K. et al., Production of drinking water from saline water by direct contact membrane distillation (DCMD). *Journal of Industrial and Engineering Chemistry*, 2011. **17**(1): pp. 41–48.

32. Curcio, E. et al., Membrane distillation operated at high seawater concentration factors: Role of the membrane on CaCO3 scaling in presence of humic acid. *Journal of Membrane Science*, 2010. **346**(2): pp. 263–269.
33. He, F., K.K. Sirkar, and J. Gilron, Studies on scaling of membranes in desalination by direct contact membrane distillation: CaCO3 and mixed CaCO3/CaSO4 systems. *Chemical Engineering Science*, 2009. **64**(8): pp. 1844–1859.
34. He, F., K.K. Sirkar, and J. Gilron, Effects of antiscalants to mitigate membrane scaling by direct contact membrane distillation. *Journal of Membrane Science*, 2009. **345**(1–2): pp. 53–58.
35. Gryta, M., Scaling diminution by heterogeneous crystallization in a filtration element integrated with membrane distillation module. *Polish Journal of Chemical Technology*, 2009. **11**(2): pp. 60–65.
36. Gryta, M., Calcium sulphate scaling in membrane distillation process. *Chemical Papers*, 2009. **63**(2): pp. 146–151.
37. Schofield, R.W. et al., Factors affecting flux in membrane distillation. *Desalination*, 1990. **77**(Supplement C): pp. 279–294.
38. Nayar, K.G. et al., Thermophysical properties of seawater: A review and new correlations that include pressure dependence. *Desalination*, 2016. **390**: pp. 1–24.
39. Sharqawy, M.H., J.H. Lienhard, and S.M. Zubair, Thermophysical properties of seawater: A review of existing correlations and data. *Desalination and Water Treatment*, 2010. **16**(1–3): pp. 354–380.
40. Hitsov, I. et al., Modelling approaches in membrane distillation: A critical review. *Separation and Purification Technology*, 2015. **142**: pp. 48–64.
41. Khayet, M., Membranes and theoretical modeling of membrane distillation: A review. *Advances in Colloid and Interface Science*, 2011. **164**(1–2): pp. 56–88.
42. Khayet, M., A. Velazquez, and J.I. Mengual, Modelling mass transport through a porous partition: Effect of pore size distribution. *Journal of Non-Equilibrium Thermodynamics*, 2004. **29**(3): pp. 279–299.
43. Younis, R.F., *An Investigation into Fouling, Wetting and Their Mitigation Using Commercial Antiscalant in Membrane Distillation Desalination*. 2017, Qatar University (Qatar): Ann Arbor, MI. p. 177.
44. Gryta, M., Calcium sulphate scaling in membrane distillation process. *Chemical Papers*, 2009. **63**(2).
45. Gryta, M., Influence of polypropylene membrane surface porosity on the performance of membrane distillation process. *Journal of Membrane Science*, 2007. **287**(1): pp. 67–78.
46. Tow, E.W. et al., Comparison of fouling propensity between reverse osmosis, forward osmosis, and membrane distillation. *Journal of Membrane Science*, 2018. **556**: pp. 352–364.
47. Darton, T. et al., Membrane autopsy helps to provide solutions to operational problems. *Desalination*, 2004. **167**: pp. 239–245.
48. Jeong, S. et al., In-depth analyses of organic matters in a full-scale seawater desalination plant and an autopsy of reverse osmosis membrane. *Separation and Purification Technology*, 2016. **162**: pp. 171–179.
49. Vrouwenvelder, J.S. and D. van der Kooij, Diagnosis of fouling problems of NF and RO membrane installations by a quick scan. *Desalination*, 2003. **153**(1): pp. 121–124.

50. Butt, F.H., F. Rahman, and U. Baduruthamal, Hollow fine fiber vs. spiral-wound reverse osmosis desalination membranes Part 2: Membrane autopsy. *Desalination*, 1997. **109**(1): pp. 83–94.
51. Dudley, L.Y. and E.G. Darton, Membrane autopsy: A case study. *Desalination*, 1996. **105**(1): pp. 135–141.
52. Butt, F.H., F. Rahman, and U. Baduruthamal, Identification of scale deposits through membrane autopsy. *Desalination*, 1995. **101**(3): pp. 219–230.
53. Dow, N. et al., Pilot trial of membrane distillation driven by low grade waste heat: Membrane fouling and energy assessment. *Desalination*, 2016. **391**: pp. 30–42.
54. Ahmed, F., N. Hilal, and R. Hashaikeh, Electrically conductive membranes for in situ fouling detection in membrane distillation using impedance spectroscopy. *Journal of Membrane Science*, 2018. **556**: pp. 66–72.
55. Yuan Y., T.R. Lee (Eds.) Contact angle and wetting properties, in *Surface Science Techniques*, H.B. Bracco G., Editor. 2013, Springer Series in Surface Sciences, vol. 51, Springer: Berlin, Germany. pp. 3–34.
56. Burgoyne, A. and M.M. Vahdati, Direct contact membrane distillation. *Separation Science and Technology*, 2000. **35**(8): pp. 1257–1284.
57. Alkhudhiri, A., N. Darwish, and N. Hilal, Membrane distillation: A comprehensive review. *Desalination*, 2012. **287**: pp. 2–18.
58. El-Bourawi, M.S. et al., A framework for better understanding membrane distillation separation process. *Journal of Membrane Science*, 2006. **285**(1–2): pp. 4–29.
59. Franken, A.C.M. et al., Wetting criteria for the applicability of membrane distillation. *Journal of Membrane Science*, 1987. **33**(3): pp. 315–328.
60. Wang, Z. and S. Lin, Membrane fouling and wetting in membrane distillation and their mitigation by novel membranes with special wettability. *Water Research*, 2017. **112**: pp. 38–47.
61. Servi, A.T. et al., A systematic study of the impact of hydrophobicity on the wetting of MD membranes. *Journal of Membrane Science*, 2016. **520**: pp. 850–859.
62. Ge, J. et al., Membrane fouling and wetting in a DCMD process for RO brine concentration. *Desalination*, 2014. **344**: pp. 97–107.
63. Bormashenko, E., Progress in understanding wetting transitions on rough surfaces. *Colloids and Surfaces A: Physicochemical and Engineering Aspects*, 2008. **324**(1): pp. 47–50.
64. Pena, L., J.M.O. Dezarate, and J.I. Mengual, Steady-states in membrane distillation: Influence of membrane wetting. *Journal of the Chemical Society-Faraday Transactions*, 1993. **89**(24): pp. 4333–4338.
65. Chen, Y. et al., Probing pore wetting in membrane distillation using impedance: Early detection and mechanism of surfactant-induced wetting. *Environmental Science & Technology Letters*, 2017. **4**(11): pp. 505–510.
66. Gryta, M., Polyphosphates used for membrane scaling inhibition during water desalination by membrane distillation. *Desalination*, 2012. **285**: pp. 170–176.
67. Gryta, M., Water desalination using membrane distillation with acidic stabilization of scaling layer thickness. *Desalination*, 2015. **365**: pp. 160–166.
68. Chen, G. et al., Performance enhancement and scaling control with gas bubbling in direct contact membrane distillation. *Desalination*, 2013. **308**: pp. 47–55.
69. Hou, D. et al., Ultrasonic assisted direct contact membrane distillation hybrid process for membrane scaling mitigation. *Desalination*, 2015. **375**: pp. 33–39.

7

Membrane Improvement in Membrane Distillation

7.1 Introduction

Membranes for membrane distillation were covered in Chapter 3. It is worth remembering that until recently, much of the research work on membrane distillation was conducted on commercially available hydrophobic microfiltration membranes (made of PTFE, PVDF and PP) with limited range physical properties (nominal pore size, porosity, thickness and support material) [1–3]. As well as using commercially available hydrophobic membranes, membranes made of the same material (PTFE, PVDF and PP) and sometimes combined with other material, were also made in the laboratory for characterization [4–16] and testing for MD application [8–13,15–19]. Desirable performance indicators for good MD membranes include high permeate flux, low heat loss, fouling resistance and wetting resistance with high quality permeate [2,3,20–24]. One more desirable membrane property seldom mentioned in the literature is mechanical strength and structural stability [25,26]. The desirable MD membrane characteristics quoted by Schneider et al. [24] in 1988 have not changed much since. Schneider et al. also stated that MD would not compete fairly against established thermal or reverse osmosis desalination on a large production volume basis but MD would establish itself in a "niche" market for ultra-pure distillate using low-grade heat or solar power. Again, their assessment is as true today as it was in 1988.

However, the reasons for the long delay in the overdue commercial deployment of MD for desalination include serious issues associated with the membrane stability and performance, specifically membrane fouling/scaling and wetting [27–29]. Without a stable high enough distillate flux and a good quality product, the viability of MD as an emerging desalination technology potentially exploiting the vast amount of dissipated low-grade heat in industry and the free, plentiful solar energy in water-stressed regions of the world will be questionable. In this chapter we will build on the lessons learnt from Chapter 6 that covers fouling and scaling, and study recent advances made in enhanced membranes for membrane distillation.

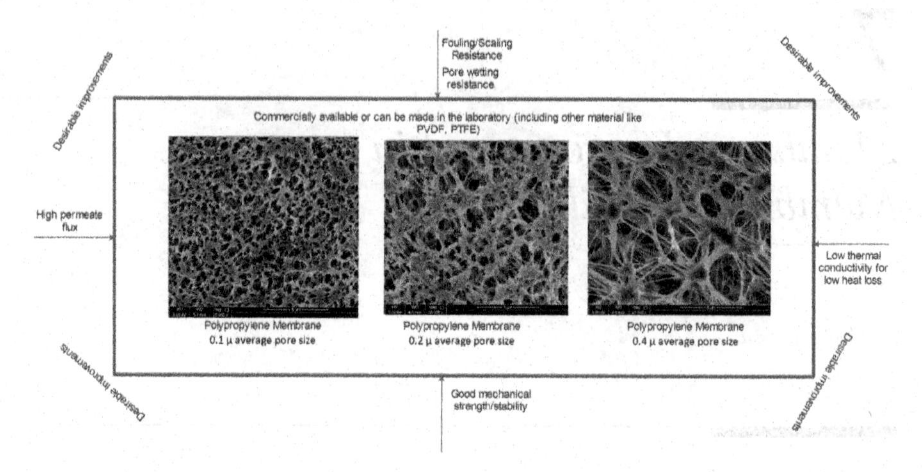

FIGURE 7.1
Desirable features of a high-performance MD membrane for desalination.

It is currently a very dynamic topic with publications claiming novelties on the topic [30–36]. We start by pointing out what are the desirable features of a high-performance MD membrane. This is depicted in Figure 7.1. At first sight it would seem paradoxical to reconcile all of these desirable attributes because some of them would negate others. For example, a high flux membrane will foul quicker. How can we synthesize a membrane that will resist fouling and have all other desirable properties?

In a recent review paper, Eykens et al. [3] indicated that an optimized MD membrane should have the "right contact and angle and liquid entry pressure (LEP)" for wetting resistance and the "right pore size, porosity, tortuosity and thermal conductivity" for high enough flux and energy efficiency. The mechanical strength was also mentioned as a separate criterion. They recommended a pore diameter of 0.3 μm to balance the liquid entry pressure with a good permeate flux. The 0.3 μm pore size was selected because out of all MD configurations, vacuum membrane distillation is more sensitive to wetting. Eykens et al. [3] also recommended an optimum membrane thickness between 10 and 700 μm, depending on process conditions to balance permeate vapor transport and thermal energy loss. The membrane porosity should also be as high as possible (greater than 75%), to improve the mass transfer and energy efficiency, while low tortuosity (1.1–1.2) and thermal conductivity (less than 0.06 $w \cdot m^{-1} \cdot K^{-1}$) were recommended. This review paper provides some sort of quantitative order of magnitude to desirable properties highlighted in Figure 7.1 but not how these high-performance membranes can be made.

Eykens et al. [2] published recently a review paper on MD membrane synthesis where they described traditional MD membrane production methods (stretching, phase inversion, sintering, track-etching and template leaching), stating that among these, stretching and phase inversion yielded higher porosity, which is highly desirable in MD. Eykens et al. [2] also described the synthesis of what they called "dedicated" MD membranes. These methods included mainly electrospinning [37–44] where nanofibers are produced to make high porosity MD membranes, and nanofiber membranes with immobilized carbon nanotubes (CNT) [45–48], emphasizing that the immobilization of CNT is still a challenge. Surface modification was mentioned in Eykens et al. [2] review and concerned the possibility to introduce surface functionality to render hydrophilic membrane material as hydrophobic. However, the flux values given in some example references were not high enough compared with traditional MD membranes. Finally, a new class of material, referred to as "novel" material with wetting resistance, was described. These materials were: ECTFE: poly(ethene-co-chlorotrifluoroethene), FEP: poly(vinylidene fluoride-co-chlorotrifluoroethylene), PVDF: polyvinylidene fluoride, PVDF-co-CTFE: poly(vinylidenefluoride-co-chlorotrifluoroethylene), PVDF-co-HFP: poly(vinylidene fluoride-co-hexafluoropropylene), PVDF-co-TFE: poly(vinylidene fluoride-co-tetrafluoroethylene) [7,49–53]. They were used to make flat sheet as well as hollow fiber membranes. Among the membrane modification techniques, "plasma" technology seems to yield membranes with quite good vapor permeability. In the plasma technology, hydrophobic coatings can be applied on polymeric surfaces [54,55] and radicals, fragmented molecules are created and polymerized with each other and the membrane surface forming a hydrophobic coating [56,57]. Drioli et al. [58] in their interesting paper entitled "Membrane distillation: Recent developments and perspectives" provided a good coverage on the electrospun nanofiber membranes with high contact angle and the potential of various modifications applied to ceramic membranes for MD applications.

In the following sections we will explore an extract of the substantial literature on enhanced and new types of MD membranes that were claimed to promote hydrophobicity and wetting resistance.

7.2 Membrane Material and Surface Modifications

7.2.1 Enhancing Membrane Flux

Zheng et al. [51] reported that hydrophobic flat-sheet membranes were prepared by polyvinylidene fluoride-co-chlorotrifluoroethylene (PVDF-CTFE) using the non-solvent induced phase separation (NIPS) process in order to test for membrane distillation (MD) application. They indicated that different types of LiCl-based mixed additives were applied to investigate their effects on membrane properties and membrane distillation performance. The membranes were evaluated in terms of membrane morphology, pore size and distribution, porosity, surface roughness, hydrophobicity, as well as the direct contact membrane distillation (DCMD) flux and permeate quality performance [51]. Their results showed that by carefully tuning the additive ratio, the hydrophobic PVDF-CTFE membrane has good prospect in hydrophobic membrane preparation, which can be prepared with optimal structure suitable for MD process. However, their membranes contact angle hardly exceeded 102° (for water) but they managed to reduce the extent of macro voids (mean pore size of 0.1 μm) and record good permeate fluxes ranging from 17 to 22 kg/(m^2 h). This is a positive outcome in terms making membranes with a narrow pore size distribution.

Huang et al. [59] fabricated polytetrafluoroethylene (PTFE) membranes with unique structures from a mixture of a PTFE emulsion and a poly(vinyl alcohol) (PVA) aqueous solution. The formation of the micro–nano structures in the PTFE membrane was due to PTFE crystallization. By controlling the cooling rate during membrane formation, various PTFE membranes with different structures and properties were prepared and showed a super-hydrophobicity with a water static contact angle of 155°. They reported an enhanced permeate flux of almost 17 kg/(m^2 h) compared to a flux of 12.60 kg/(m^2 h) for a standard PTFE membrane using vacuum membrane distillation module. However, their membrane had a lower porosity (45.8%) but higher pore size (0.68 μm) compared to 68.3% and 0.39 μm for the standard PTFE membrane. They used cold stretching to produce the pores in their new membranes.

Leaper et al. [60] described in their study the fabrication of high flux robust membranes for the purification of synthetic seawater by incorporating graphene oxide functionalized with 3-(aminopropyl)triethoxysilane (APTS) into PVDF polymer solutions. They confirmed successful functionalization of GO with APTS using XPS and FTIR analytical tools. They showed that the addition of GO and GO-APTS enhanced the permeate flux by 52%–86%, respectively, compared to pure PVDF. They used an air gap MD unit with a feed temperature of 80°C. Their membranes had thicknesses ranging from 19.9–25.1 μm. They also claimed that the best performing membrane contained 0.3 wt% GO-APTS (with respect to PVDF) and had a flux of 6.2 Lm^{-2} h^{-1}, while maintaining high salt rejection (greater than 99.9%). They attributed the improvements to increased surface and bulk porosity, larger mean pore size and hydrophilic interactions owing to the functional groups of GO and GO-APTS. These membranes were claimed to be evidence of the potential that GO and related materials have as nanocomposite fillers in high performance desalination membranes.

7.2.2 Enhanced Membrane Hydrophobicity and Wetting Resistance

The membrane hydrophobicity is one of the most important properties of membrane distillation membranes. Ahmad et al. [61] gave an account on membrane hydrophobicity and how hydrophobicity can be measured experimentally using a variety of tools such as contact angle and atomic force microscopy. They also described literature methods on how to enhance the membrane hydrophobicity by blending, additives and surface modifications.

Pore wetting is an important membrane failure indicator in membrane distillation. The existing approach to membrane wetting detection is currently based on distillate conductivity measurements and works only when a membrane has fouled and its pores become wetted. Chen et al. [62] developed a method, based on measurement of cross-membrane impedance, for monitoring the dynamics of membrane pore wetting and enabling early detection of imminent wetting-based membrane failure. Using non-anionic surfactant Triton X-100 to induce pore wetting in direct contact membrane distillation experiments (using a commercial PVDF membrane with 0.45 and 175 μm average pore size and membrane thickness respectively with a feed consisting of 0.6 M NaCl solution and 60/20°C temperatures in a direct contact MD system), they demonstrated the rapid response of single-frequency impedance to partial pore wetting long before any change in distillate conductivity was observed. Chen et al. [62] also conducted membrane distillation experiments using alternating feed solutions with and without surfactants to elucidate the mechanism of surfactant-induced pore wetting. Their experimental observations suggested that surfactant-induced pore wetting occurred via progressive movement of the water–air interface and that adsorption of surfactants to the membrane pore surface plays an important role in controlling the kinetics of progressive wetting. The work of Chen et al. [62] is an important contribution toward understanding the membrane wetting mechanism and also obtaining an early warning indicator that the membrane will fail with deteriorated permeate quality.

7.2.3 Enhanced Mechanical Properties

Zhao et al. [14] developed polyvinylidene fluoride (PVDF)/polytetrafluoroethylene (PTFE) hollow fiber membranes via thermally induced phase separation (TIPS) method for direct contact membrane distillation application. The effects of PTFE addition on the thermal behavior of the dope mixtures and membrane formation were investigated. Zhao et al. found that the crystallization of PVDF was significantly enhanced with increased nucleation sites provided by PTFE particles, leading to promoted formation of smaller spherulites in a greater density. In addition, the improved uniformity and increased amount of cavity between the spherical crystallites enabled the formation of smaller pores ranging from 0.08 to 0.12 μm. The membranes were tested for DCMD performance under various conditions using NaCl solution (3.5 wt %). A stable permeate flux of 28.3 kg/(m^2 h) for a feed temperature of 60°C with 99.99% NaCl rejection for over 50 h of operation was reported. According to Zhao et al. this performance is comparable to similar type of PVDF membranes while the newly developed membrane exhibited better mechanical strength (elongation and tensile strength) [14].

7.3 New and Novel Membrane Distillation Membranes

Yang et al. [63] argued that metal–organic frameworks (MOFs) are gaining importance as fillers for desalination composite membranes. In their study, superhydrophobic poly(vinylidene fluoride) nanofibrous membranes were fabricated with MOF (iron 1,3,5-benzenetricarboxylate) loading of up to 5 wt% via electrospinning on a nonwoven substrate. To improve the attachment of nanofibers onto the substrate, a substrate pretreatment method called "solvent basing" was employed. The iron content in the nanofiber, measured by energy dispersive X-ray spectroscopy, increased proportionally with the increase of the MOF concentration in the spinning dope, indicating a uniform distribution of MOF in the nanofiber. The water contact angle was claimed to have increased up to 138.06° upon incorporation of 5 wt% MOF, and a liquid entry pressure of 82.73 kPa could be maintained, making the membrane useful for direct contact membrane distillation experiments. The membrane was reported to be stable for a period of 5 h, exhibiting a water vapor flux of 2.87 $kg/m^2 \cdot h$ and 99.99% salt rejection (NaCl concentration of 35 g/L) when the feed and permeate temperatures were 48°C and 16°C, respectively. One can argue that Yang et al. [63] experiments were conducted under mildly fouling conditions and that further MD tests should be undertaken at higher feed temperatures where the flux will be much higher and the potential for membrane fouling may also be significantly increased. That would be the ultimate test to prove that the membrane has good fouling resistance.

Qiu et al. [30] described a membrane surface modification by forming a functional layer that is an effective way to improve antifouling properties of membranes. They argued that additional layer and the potential blockage of bulk pores may increase the mass transfer resistance and reduce the permeability. To address this study, they applied a novel method of preparing anti-fouling membranes for membrane distillation by dispersing graphene oxide (GO) on the channel surface of polyvinylidene fluoride membranes. The surface morphology and properties were characterized by scanning electron microscopy, atomic force microscope and Fourier transform infrared spectrometry.

The literature provides a great deal of additional sources on modified MD membranes that have potentially beneficial application in membrane distillation [64–83].

7.4 Omniphobic and Amphiphobic Membranes

A material is omniphobic when it is simultaneously repellent to both water and oil [84]. This is a remarkable property that is attracting a growing interest in the field of new membrane synthesis for membrane distillation applications where wetting is still an outstanding issue. Composite omniphobic membranes were synthesized, tested in MD and were claimed to be suitable for future treatment of saline wastewater containing low surface tension organic contaminants [84].

Woo et al. [85] carried out work in which composite membranes were fabricated via layer-by-layer (LBL) assembly of negatively charged silica aerogel (SiA) and 1H,1H,2H,2H-perfluorodecyltriethoxysilane (FTCS) on a polyvinylidene fluoride phase inversion membrane and interconnecting them with positively charged poly(diallyldimethylammonium chloride) (PDDA) via electrostatic interaction. Their experimental results showed that the PDDA-SiA-FTCS coated membrane improved the membrane structure and properties. They saw that new trifluoromethyl and tetrafluoroethylene bonds appeared at the surface of the coated membrane leading to a lower surface free energy of the composite membrane. Additionally, the LBL membrane appeared to have increased its surface roughness. The improved structure and property gave the LBL membrane an omniphobic property, as evidenced by its good wetting resistance. The membrane was tested and showed a stable performance in an air gap membrane distillation (AGMD) unit with a flux of 11.22 L/(m^2 h) with very high salt rejection using reverse osmosis brine from coal seam gas produced water as feed with the addition of up to 0.5 mM SDS solution.

Lu et al. [86] reported that an omniphobic polyvinylidene difluoride (PVDF) hollow-fiber membrane was developed via silica nanoparticle deposition followed by a Teflon AF 2400 coating in this study. They claimed that their membrane showed good repellency toward various liquids with different surface tensions and chemistries, including water, ethylene glycol (EG), cooking oil and ethanol. They tested the membrane in a VMD system and their membrane appeared to exhibit stable performance in 7 h tests with a feed solution containing up to 0.6 mM of sodium dodecyl sulfate (SDS). However, not enough information was provided for the MD tests. They also indicated that the effects of surface energy and surface morphology as well as nanoparticle size on membrane omniphobicity have been systematically investigated. Obviously, more work is needed under a wide range of MD conditions to ascertain the long term viability of such membranes.

Lu et al. [87] described work in a recent paper where a novel omniphobic nanofiber membrane for membrane distillation (MD) was fabricated by one-step electrospinning of PVDF-HFP and fluorinated-decyl polyhedraloligomeric silsesquioxane (F-POSS) colloidal suspension solution. The F-POSS/PVDF-HFP membrane prepared was claimed to have uniform fiber structures and very high F-POSS concentration on the surface. The F-POSS based membrane also exhibited excellent omniphobicity, due to its wetting resistance to different low surface tension liquids such as ethanol where a contact angle of 128.2° was measured. Membrane distillation (MD) experiments showed that stable water flux and water quality were obtained in highly saline feed solutions containing low surface tension substances.

This work was claimed to demonstrate the membrane potential for application in desalination of challenging industrial wastewater.

Chen et al. [88] reported work where an omniphobic membrane was fabricated for membrane distillation application by effectively depositing ZnO nanoparticles on a hydrophilic glass fiber (GF) membrane using a chemical bath deposition method to create hierarchical re-entrant structures, followed by surface fluorination and the addition of a polymer coating to lower the surface energy of the membrane. The omniphobic membranes prepared were claimed to have a particulate membrane morphology and very high fluorine concentration on the surface. The omniphobic nature of the membrane fabricated was proved by the contact angles for water and ethanol, which were measured as high as 152.8 ± 1.1° and 110.3 ± 1.9°, respectively. Compared to hydrophobic GF membranes without deposited ZnO nanoparticles, the omniphobic membrane displayed a higher wetting resistance to low surface tension feed solutions in direct membrane distillation runs. The water flux was observed to be stable in a 0.3 mM sodium dodecyl sulfate (SDS) solution (in a feed at 60°C 1M NaCl) and that with the salt rejection was higher than 99.99%. The results obtained indicated that the omniphobic membrane exhibited not only superior wetting resistance to low surface tension liquids but also a potential for desalinating low surface tension wastewater.

In a recent paper, Zheng et al. [89] described an omniphobic polyvinylidene fluoride (PVDF) membrane with hierarchical structure which was created by spray coating of the nano/microspheres onto a commercial PVDF porous substrate. They measured the contact angle of water/hexadecane present on the membrane surface and recorded values of 176°/138.4°. The membrane was tested in a MD system with a hexadecane emulsion feed solution and observed a stable flux and decrease in conductivity in the permeate side in contrast to a decrease in flux and increase in permeate conductivity for other benchmark membranes. Their work showed that omniphobic membranes with hierarchical morphology are promising in addressing membrane wetting and fouling issues in the DCMD.

Amphiphobic surfaces have high surface roughness and low surface energy, which is beneficial for completely repelling low surface tension liquids. An et al. [90] reported work on a simple and facile approach to fabricate amphiphobic polyvinylidene fluoride-co-hexafluoropropylene (PVDF-HFP) electrospun nanofibrous membranes with anti-wetting property. The perfluorodecyltriethoxysilane (FAS)-coated PVDF-HFP nanofibrous membranes showed stable amphiphobicity with high contact angles against both water and oil. Surface fluorination of PVDF-HFP nanofibrous membrane with the FAS coating did not negatively affect the permeate flux and salt rejection in a significant way. The amphiphobic nanofibrous membrane also exhibited robust MD performance in a long-term operation in the presence of surfactant sodium dodecyl sulfate in the feed saline solution.

7.5 Bioinspired MD Membranes

Zhu et al. [91] reported work where they described a method to fabricate superhydrophobic polyimide nanofibrous membranes (PI NFMs) with hierarchical structures, interconnected pores and high porosity, which was derived from the electrospinning, dual-bioinspired design and fluorination processes. The resultant superhydrophobic PI NFMs had a water contact angle of 152°, high hot water resistance up to 85°C, and entry pressure of 42 kPa. The membrane with omniphobicity displayed a high permeate flux over 31 L m^{-2} h^{-1} and high salt rejection of nearly 100% as well as robust durability for treating high salinity wastewater containing typical low surface tension and dissolved contaminants (ΔT = 40°C). More importantly, the authors claimed that the novel dual-bioinspired method can be exploited as a universal tool to alter various materials with hierarchical structures, which is expected to lead to more effective alternative membranes for membrane distillation applications.

7.6 Novel Janus Membranes

Janus membranes are a novel class of membranes with asymmetric properties on each side [34].

Huang et al. [33] reported recently a novel Janus membrane integrating an omniphobic substrate and an in-air hydrophilic, underwater superoleophobic skin layer. The new membrane was claimed to enable membrane distillation to desalinate hypersaline brine with both hydrophobic foulants and amphiphilic wetting agents. Engineered to overcome the limitations of existing MD membranes, the Janus membrane has been shown to exhibit novel wetting properties not to be found in any existing membrane, including hydrophobic membranes, omniphobic membranes and hydrophobic membranes with a hydrophilic surface coating. The authors quoted "being simultaneously resistant to both membrane fouling and wetting, a Janus membrane can sustain stable MD performance even with challenging feed waters and can thus potentially transform MD to be a viable technology for desalinating hypersaline wastewater with complex compositions using low-grade-thermal energy."

7.7 Concluding Remarks

The message for the need for improved MB membranes was certainly received by the scientific community. This has been echoed with a tremendous surge in new membrane development work. In this chapter we have seen the diversity of channels taken to come up with membranes that have superhydrophobic properties, wetting resistance properties and sometimes mechanical strength. These channels included surface modification of existing MD membranes, new co-polymer membranes, nanofiber membranes, immobilized carbon nanotube membranes, omniphobic and amphiphobic membranes and Janus membranes. With such diversity, there is certainly choice. However, in the end, only classic MD experiments using real seawater feed and temperatures typically encountered using solar power or low-grade waste heat will determine the wetting resistance of these new membranes. Unfortunately, despite the huge number of papers in this emerging topic, we have yet to see a new membrane that matches the gold standard of thin film composite membranes in reverse osmosis. The final message is: there is need for a thorough testing of any new MD membrane under realistic desalination conditions to achieve the aim, namely a MD membrane that is wetting resistant and has permeate high flux with low thermal conductivity and good mechanical strength.

References

1. Wang, P. and T.S. Chung, Recent advances in membrane distillation processes: Membrane development, configuration design and application exploring. *Journal of Membrane Science*, 2015. **474**: pp. 39–56.
2. Eykens, L. et al., Membrane synthesis for membrane distillation: A review. *Separation and Purification Technology*, 2017. **182**: pp. 36–51.
3. Eykens, L. et al., How to optimize the membrane properties for membrane distillation: A review. *Industrial & Engineering Chemistry Research*, 2016. **55**(35): pp. 9333–9343.
4. Kitamura, T. et al., Formation mechanism of porous structure in polytetrafluoroethylene (PTFE) porous membrane through mechanical operations. *Polymer Engineering & Science*, 1999. **39**(11): pp. 2256–2263.
5. Kong, L.F. and K. Li, Preparation of PVDF hollow-fiber membranes via immersion precipitation. *Journal of Applied Polymer Science*, 2001. **81**(7): pp. 1643–1653.
6. Chandavasu, C. et al., Polypropylene blends with potential as materials for microporous membranes formed by melt processing. *Polymer*, 2002. **43**(3): pp. 781–795.
7. Feng, C.S. et al., Preparation and properties of microporous membrane from poly(vinylidene fluoride-co-tetrafluoroethylene) (F2.4) for membrane distillation. *Journal of Membrane Science*, 2004. **237**(1–2): pp. 15–24.
8. Huo, R. et al., Preparation and properties of PVDF-fabric composite membrane for membrane distillation.*Desalination*, 2009. **249**(3): pp. 910–913.
9. Simone, S. et al., Preparation of hollow fibre membranes from PVDF/PVP blends and their application in VMD. *Journal of Membrane Science*, 2010. **364**(1–2): pp. 219–232.

10. Lai, C.L. et al., Preparation and characterization of plasma-modified PTFE membrane and its application in direct contact membrane distillation. *Desalination*, 2011. **267**(2–3): pp. 184–192.
11. Zhu, H.L. et al., Preparation and properties of PTFE hollow fiber membranes for desalination through vacuum membrane distillation. *Journal of Membrane Science*, 2013. **446**: pp. 145–153.
12. Hou, D.Y. et al., Preparation and characterization of PVDF flat-sheet membranes for direct contact membrane distillation. *Separation and Purification Technology*, 2014. **135**: pp. 211–222.
13. Xiao, T.H. et al., Fabrication and characterization of novel asymmetric polyvinylidene fluoride (PVDF) membranes by the nonsolvent thermally induced phase separation (NTIPS) method for membrane distillation applications. *Journal of Membrane Science*, 2015. **489**: pp. 160–174.
14. Zhao, J. et al., Preparation of PVDF/PTFE hollow fiber membranes for direct contact membrane distillation via thermally induced phase separation method. *Desalination*, 2018. **430**: pp. 86–97.
15. Dong, Z.-Q. et al., Superhydrophobic PVDF–PTFE electrospun nanofibrous membranes for desalination by vacuum membrane distillation. *Desalination*, 2014. **347**: pp. 175–183.
16. Bonyadi, S. and T.-S. Chung, Highly porous and macrovoid-free PVDF hollow fiber membranes for membrane distillation by a solvent-dope solution co-extrusion approach. *Journal of Membrane Science*, 2009. **331**(1–2): pp. 66–74.
17. Tomaszewska, M., Preparation and properties of flat-sheet membranes from poly(vinylidene fluoride) for membrane distillation. *Desalination*, 1996. **104**(1–2): pp. 1–11.
18. Khayet, M. and T. Matsuura, Preparation and characterization of polyvinylidene fluoride membranes for membrane distillation. *Industrial & Engineering Chemistry Research*, 2001. **40**(24): pp. 5710–5718.
19. Wu, B., K. Li, and W.K. Te, Preparation and characterization of poly(vinylidene fluoride) hollow fiber membranes for vacuum membrane distillation. *Journal of Applied Polymer Science*, 2007. **106**(3): pp. 1482–1495.
20. Luo, A. and N. Lior, Critical review of membrane distillation performance criteria. *Desalination and Water Treatment*, 2016. **57**(43): pp. 20093–20140.
21. Alkhudhiri, A., N. Darwish, and N. Hilal, Membrane distillation: A comprehensive review. *Desalination*, 2012. **287**: pp. 2–18.
22. El-Bourawi, M.S. et al., A framework for better understanding membrane distillation separation process. *Journal of Membrane Science*, 2006. **285**(1–2): pp. 4–29.
23. Khayet, M., Membranes and theoretical modeling of membrane distillation: A review. *Advances in Colloid and Interface Science*, 2011. **164**(1–2): pp. 56–88.
24. Schneider, K. et al., Membranes and modules for transmembrane distillation. *Journal of Membrane Science*, 1988. **39**(1): pp. 25–42.
25. Lawson, K.W., M.S. Hall, and D.R. Lloyd, Compaction of microporous membranes used in membrane distillation. I. Effect on gas permeability. *Journal of Membrane Science*, 1995. **101**(1): pp. 99–108.
26. Rao, G. et al., Factors contributing to flux improvement in vacuum-enhanced direct contact membrane distillation. *Desalination*, 2015. **367**: pp. 197–205.
27. Naidu, G. et al., A review on fouling of membrane distillation. *Desalination and Water Treatment*, 2016. **57**(22): pp. 10052–10076.

28. Tijing, L.D. et al., Fouling and its control in membrane distillation—A review. *Journal of Membrane Science*, 2015. **475**: pp. 215–244.
29. Gryta, M., Fouling in direct contact membrane distillation process. *Journal of Membrane Science*, 2008. **325**(1): pp. 383–394.
30. Qiu, H. et al., Pore channel surface modification for enhancing anti-fouling membrane distillation. *Applied Surface Science*, 2018. **443**: pp. 217–226.
31. Zuo, J. and T.-S. Chung, PVDF hollow fibers with novel sandwich structure and superior wetting resistance for vacuum membrane distillation. *Desalination*, 2017. **417**: pp. 94–101.
32. Wang, Z. and S. Lin, Membrane fouling and wetting in membrane distillation and their mitigation by novel membranes with special wettability. *Water Research*, 2017. **112**: pp. 38–47.
33. Huang, Y.-X. et al., Novel janus membrane for membrane distillation with simultaneous fouling and wetting resistance. *Environmental Science & Technology*, 2017. **51**(22): pp. 13304–13310.
34. Yang, H.-C. et al., Janus membranes: Creating asymmetry for energy efficiency. *Advanced Materials*, 2018. **30**(43): p. 1801495.
35. Yang, H.-C. et al., Janus hollow fiber membrane with a mussel-inspired coating on the lumen surface for direct contact membrane distillation. *Journal of Membrane Science*, 2017. **523**: pp. 1–7.
36. Bhadra, M., S. Roy, and S. Mitra, Desalination across a graphene oxide membrane via direct contact membrane distillation. *Desalination*, 2016. **378**: pp. 37–43.
37. Ray, S.S. et al., A comprehensive review: Electrospinning technique for fabrication and surface modification of membranes for water treatment application. *RSC Advances*, 2016. **6**(88): pp. 85495–85514.
38. Brown, T.D., P.D. Dalton, and D.W. Hutmacher, Melt electrospinning today: An opportune time for an emerging polymer process. *Progress in Polymer Science*, 2016. **56**: pp. 116–166.
39. Agarwal, S., M. Burgard, A. Greiner, and J. Wendorff. *Electrospinning: A Practical guide to Nanofibers*. 2016, Walter de Gruyter GmbH & Co KG.
40. Mitchell, G.R., *Electrospinning: Principles, Practice and Possibilities*. 2015, Royal Society of Chemistry.
41. Liu, Z. et al., Tug of war effect in melt electrospinning. *Journal of Non-Newtonian Fluid Mechanics*, 2013. **202**: pp. 131–136.
42. Essalhi, M. and M. Khayet, Self-sustained webs of polyvinylidene fluoride electrospun nanofibers at different electrospinning times: 2. Theoretical analysis, polarization effects and thermal efficiency. *Journal of Membrane Science*, 2013. **433**: pp. 180–191.
43. Bhardwaj, N. and S.C. Kundu, Electrospinning: A fascinating fiber fabrication technique. *Biotechnology Advances*, 2010. **28**(3): pp. 325–347.
44. Lyons, J., C. Li, and F. Ko, Melt-electrospinning part I: Processing parameters and geometric properties. *Polymer*, 2004. **45**(22): pp. 7597–7603.
45. Tijing, L.D. et al., Superhydrophobic nanofiber membrane containing carbon nanotubes for high-performance direct contact membrane distillation. *Journal of Membrane Science*, 2016. **502**: pp. 158–170.
46. Lee, J.-G. et al., Theoretical modeling and experimental validation of transport and separation properties of carbon nanotube electrospun membrane distillation. *Journal of Membrane Science*, 2017. **526**: pp. 395–408.

47. Dumee, L.F. et al., Characterization and evaluation of carbon nanotube bucky-paper membranes for direct contact membrane distillation. *Journal of Membrane Science*, 2010. **351**(1–2): pp. 36–43.
48. Roy, S., M. Bhadra, and S. Mitra, Enhanced desalination via functionalized carbon nanotube immobilized membrane in direct contact membrane distillation. *Separation and Purification Technology*, 2014. **136**: pp. 58–65.
49. Gryta, M., The study of performance of polyethylene chlorinetrifluoroethylene membranes used for brine desalination by membrane distillation. *Desalination*, 2016. **398**: pp. 52–63.
50. Pan, J. et al., ECTFE porous membranes with conveniently controlled microstructures for vacuum membrane distillation. *Journal of Materials Chemistry A*, 2015. **3**(46): pp. 23549–23559.
51. Zheng, L. et al., Preparation of PVDF-CTFE hydrophobic membranes for MD application: Effect of LiCl-based mixed additives. *Journal of Membrane Science*, 2016. **506**: pp. 71–85.
52. Chen, K.K. et al., Study on vacuum membrane distillation (VMD) using FEP hollow fiber membrane. *Desalination*, 2015. **375**: pp. 24–32.
53. Khayet, M. et al., Hollow fiber spinning experimental design and analysis of defects for fabrication of optimized membranes for membrane distillation. *Desalination*, 2012. **287**: pp. 146–158.
54. Inagaki, N. and J. Ohkubo, Plasma polymerization of hexafluoropropene/methane mixtures and composite membranes for gas separations. *Journal of Membrane Science*, 1986. **27**(1): pp. 63–75.
55. Yasuda, H., Plasma polymerization for protective coatings and composite membranes. *Journal of Membrane Science*, 1984. **18**: pp. 273–284.
56. Wu, Y. et al., Surface-modified hydrophilic membranes in membrane distillation. *Journal of Membrane Science*, 1992. **72**(2): pp. 189–196.
57. Jin, Z. et al., Hydrophobic modification of poly(phthalazinone ether sulfone ketone) hollow fiber membrane for vacuum membrane distillation. *Chinese Chemical Letters*, 2008. **19**(3): pp. 367–370.
58. Drioli, E., A. Ali, and F. Macedonio, Membrane distillation: Recent developments and perspectives. *Desalination*, 2015. **356**: pp. 56–84.
59. Huang, Q.-L. et al., Design of super-hydrophobic microporous polytetrafluoroethylene membranes. *New Journal of Chemistry*, 2013. **37**(2): pp. 373–379.
60. Leaper, S. et al., Flux-enhanced PVDF mixed matrix membranes incorporating APTS-functionalized graphene oxide for membrane distillation. *Journal of Membrane Science*, 2018. **554**: pp. 309–323.
61. Ahmad, N.A. et al., Membranes with great hydrophobicity: A review on preparation and characterization. *Separation and Purification Reviews*, 2015. **44**(2): pp. 109–134.
62. Chen, Y. et al., Probing pore wetting in membrane distillation using impedance: Early detection and mechanism of surfactant-induced wetting. *Environmental Science & Technology Letters*, 2017. **4**(11): pp. 505–510.
63. Yang, F. et al., Metal–Organic rrameworks supported on nanofiber for desalination by direct contact membrane distillation. *ACS Applied Materials & Interfaces*, 2018. **10**(13): pp. 11251–11260.
64. Eykens, L. et al., Coating techniques for membrane distillation: An experimental assessment. *Separation and Purification Technology*, 2018. **193**: pp. 38–48.

65. Eykens, L. et al., Atmospheric plasma coatings for membrane distillation. *Journal of Membrane Science*, 2018. **554**: pp. 175–183.
66. Bhadra, M., S. Roy, and S. Mitra, Flux enhancement in direct contact membrane distillation by implementing carbon nanotube immobilized PTFE membrane. *Separation and Purification Technology*, 2016. **161**: pp. 136–143.
67. Duong, H.C. et al., A novel electrospun, hydrophobic, and elastomeric styrene-butadiene-styrene membrane for membrane distillation applications. *Journal of Membrane Science*, 2018. **549**: pp. 420–427.
68. Zheng, L. et al., Preparation, evaluation and modification of PVDF-CTFE hydrophobic membrane for MD desalination application. *Desalination*, 2017. **402**: pp. 162–172.
69. Zhao, D. et al., Fluorographite modified PVDF membranes for seawater desalination via direct contact membrane distillation. *Desalination*, 2017. **413**: pp. 119–126.
70. Gethard, K., O. Sae-Khow, and S. Mitra, Carbon nanotube enhanced membrane distillation for simultaneous generation of pure water and concentrating pharmaceutical waste. *Separation and Purification Technology*, 2012. **90**: pp. 239–245.
71. Ragunath, S., S. Roy, and S. Mitra, Carbon nanotube immobilized membrane with controlled nanotube incorporation via phase inversion polymerization for membrane distillation based desalination. *Separation and Purification Technology*, 2018. **194**: pp. 249–255.
72. Dumee, L. et al., Enhanced durability and hydrophobicity of carbon nanotube bucky paper membranes in membrane distillation. *Journal of Membrane Science*, 2011. **376**(1–2): pp. 241–246.
73. Liao, Y. et al., Electrospun superhydrophobic membranes with unique structures for membrane distillation. *ACS Applied Materials & Interfaces*, 2014. **6**(18): pp. 16035–16048.
74. Mejia Mendez, D.L. et al., Membrane distillation (MD) processes for water desalination applications. Can dense selfstanding membranes compete with microporous hydrophobic materials? *Chemical Engineering Science*, 2018. **188**: pp. 84–96.
75. Woo, Y.C. et al., Electrospun dual-layer nonwoven membrane for desalination by air gap membrane distillation. *Desalination*, 2017. **403**: pp. 187–198.
76. Li, X. et al., Electrospun superhydrophobic organic/inorganic composite nanofibrous membranes for membrane distillation. *ACS Applied Materials & Interfaces*, 2015. **7**(39): pp. 21919–21930.
77. Tang, N. et al., Preparation of a hydrophobically enhanced antifouling isotactic polypropylene/silicone dioxide flat-sheet membrane via thermally induced phase separation for vacuum membrane distillation. *Journal of Applied Polymer Science*, 2015. **132**(40).
78. Zhang, W. et al., Fabrication of hierarchical poly (vinylidene fluoride) micro/nano-composite membrane with anti-fouling property for membrane distillation. *Journal of Membrane Science*, 2017. **535**: pp. 258–267.
79. Ahmed, F.E., B.S. Lalia, and R. Hashaikeh, A review on electrospinning for membrane fabrication: Challenges and applications. *Desalination*, 2015. **356**: pp. 15–30.
80. Zhang, Y. et al., Enhancing wetting resistance of poly(vinylidene fluoride) membranes for vacuum membrane distillation. *Desalination*, 2017. **415**: pp. 58–66.

81. Hammami, M.A. et al., Engineering hydrophobic organosilica nanoparticle-doped nanofibers for enhanced and fouling resistant membrane distillation. *ACS Applied Materials & Interfaces*, 2017. **9**(2): pp. 1737–1745.
82. Lee, E.-J. et al., Engineering the re-entrant hierarchy and surface energy of PDMS-PVDF membrane for membrane distillation using a facile and benign microsphere coating. *Environmental Science & Technology*, 2017. **51**(17): pp. 10117–10126.
83. Guo, F. et al., Desalination by membrane distillation using electrospun polyamide fiber membranes with surface fluorination by chemical vapor deposition. *ACS Applied Materials & Interfaces*, 2015. **7**(15): pp. 8225–8232.
84. Wang, Z., M. Elimelech, and S. Lin, Environmental applications of interfacial materials with special wettability. *Environmental Science & Technology*, 2016. **50**(5): pp. 2132–2150.
85. Woo, Y.C. et al., Hierarchical composite membranes with robust omniphobic surface using layer-by-layer assembly technique. *Environmental Science & Technology*, 2018. **52**(4): pp. 2186–2196.
86. Lu, K.J. et al., Omniphobic hollow-fiber membranes for vacuum membrane distillation. *Environmental Science & Technology*, 2018. **52**(7): pp. 4472–4480.
87. Lu, C. et al., F-POSS based omniphobic membrane for robust membrane distillation. *Materials Letters*, 2018. **228**: pp. 85–88.
88. Chen, L.-H. et al., Omniphobic membranes for direct contact membrane distillation: Effective deposition of zinc oxide nanoparticles. *Desalination*, 2018. **428**: pp. 255–263.
89. Zheng, R. et al., Preparation of omniphobic PVDF membrane with hierarchical structure for treating saline oily wastewater using direct contact membrane distillation. *Journal of Membrane Science*, 2018. **555**: pp. 197–205.
90. An, X., Z. Liu, and Y. Hu, Amphiphobic surface modification of electrospun nanofibrous membranes for anti-wetting performance in membrane distillation. *Desalination*, 2018. **432**: pp. 23–31.
91. Zhu, Z. et al., Dual-Bioinspired design for constructing membranes with superhydrophobicity for direct contact membrane distillation. *Environmental Science & Technology*, 2018. **52**(5): pp. 3027–3036.

8

Modeling of Membrane Distillation

8.1 Introduction

Membrane distillation is one of the most researched separation technology topics in the past couple of decades for a variety of applications ranging from wastewater treatment and desalination to fruit juice concentration. It is also a research topic studied by researchers from a great variety of backgrounds. This is quite significant for a variety of reasons, including impact of dissemination of results in a wide range of journals foci and conferences.

In order not to confuse this emerging technology with a "similar sounding technology," namely pervaporation [1,2], we recall from Chapter 1 the outcome of the membrane distillation (MD) nomenclature meeting that yielded some very important original characteristics that give MD a distinctive identity [3]:

1. The membrane should be porous;
2. The membrane should not be wetted by process liquids;
3. No capillary condensation should take place inside the pores of the membrane;
4. Only vapor should be transported through the pores of the membrane;
5. The membrane must not alter the vapor equilibrium of the different components in the process liquids;
6. At least one side of the membrane should be in direct contact with the process liquid; and
7. For each component, the driving force of the membrane operation is a partial pressure gradient in the vapor phase.

In order to describe mathematically the MD system behavior and making sure that the original characteristics of MD are adhered to, one can envisage a number of items that must be considered. Taking one of the most popular MD configurations as an example, namely the "direct contact membrane distillation" or DCMD, we can initiate the steps that lead to a set of equations with associated boundary conditions, system components properties correlations and simplifying assumptions, that constitute a system model. Obviously, this approach is not the only one available, and other ways to model MD will be covered later in this chapter.

Before we start describing the "model building steps," it is instructive to remind ourselves the motivation to model the MD system (or for that matter, any system) and what are the benefits.

The literature quotes a number of ways a mathematical model can be defined, but we can say in very simple terms that a mathematical model of a system is a set of equations and associated concepts, which, when solved, provide useful information about the system behavior. The main benefits of a system mathematical model include understanding the system behavior when parameters are varied, improving the system performance through mathematical optimization (optimization is a higher level of the model) and assisting in the design of the system components through reduction of some unnecessary and costly practical experimentations. This is by no means an exhaustive list of benefits of mathematical modeling.

Let us now briefly review the DCMD system (for desalination) and its components (more details were provided in Chapters 2 and 4): the distillation cell is enclosed in a module compartment, and for a simple geometry like the flat sheet membrane, as depicted in Figures 4.6, 5.1 and 8.1, we can clearly see three system components: the hot feed side channel and its fluid, the separation hydrophobic microporous membrane and the cold side channel (also known as permeate channel) and its fluid. These three distinct system components have occasionally been referred to as zones (such as in Figure 5.1 in Chapter 5) or regions (region 1, region 2, region 3) for convenient referral [4]. For a desalination operation, the hot feed fluid will be hot brine (seawater, synthetic brine or concentrated industrial waste brine from various sources) and the cold fluid will be pure distillate water. Let us inspect each of these components.

The feed channel will have a geometry with its specific characteristic dimensions, and the prevailing operating conditions (temperature, flowrate) will affect the feed fluid physical properties and flow patterns (e.g., velocity, turbulence). The channel pressure can be monitored for an important purpose: to ensure its value is less than the membrane specific "liquid entry pressure" or wetting pressure, covered in Chapter 3.

The cold permeate channel has a similar approach to the feed channel for the simple flat sheet membrane module, and some operational variants include direction of flow: the common counter-current flow versus the less used co-current flow that has been reported to be less performing.

On the other hand, the separation hydrophobic microporous membrane will have specific material and morphological properties (e.g., porosity, average or "nominal" pore size) and will have one side exposed to the hot feed and the other side exposed to the cold permeate for the direct contact configuration.

Clearly, from this simple system description, we can see that the modeling challenge may well be centered around the membrane component, and that is true. Slight complications may be created in the feed and permeate channels by incorporating turbulence promoters called "spacers," which are essentially some form of "static mixers." Turbulence in MD module compartments has been historically a mildly contentious topic because of the need to apply "suitable" transfer phenomena correlations. This will be covered later in this chapter.

What about modeling more complicated MD modules such as DCMD with hollow fiber membranes, or other configurations such as vacuum membrane distillation (VMD), air gap membrane distillation (AGMD) and the comparatively less used sweeping gas membrane distillation (SGMD)?

Let us take these cases in turn. For hollow fiber DCMD modules, the concepts are virtually identical to the classic flat sheet DCMD module and only geometric considerations need to be accounted for, for both the module shell and bundle of membranes. The salty feed tends to be allocated to the shell side to reduce operational issues in case of severe scaling/fouling, and the permeate vapor flow direction within the capillary (hollow fiber) membrane tube is termed "cross flow" for obvious reasons.

In a VMD system, the feed side is virtually identical to the DCMD, and this is true for any configuration. However, the permeate side for VMD has no liquid in contact with the membrane (regardless of the shape, flat sheet or hollow fiber) and its pressure is much less than the interfacial boundary layer pressure in the feed side, approaching vacuum by means of a vacuum pump. The pressure difference between the membrane-feed interface pressure and the vacuum (low pressure) in the permeate side constitute the driving force for vapor transport. The permeate vapor is forced to condense outside the module using various types of condensers (typically using cryogenic systems for efficient permeate recovery). The vacuum created in the permeate side offers some advantages over DCMD, namely greatly reduced conductive heat transfer through the membrane and reduced mass transfer resistance of the permeate. This concept obviously has implications on modeling, as some heat transfer mechanisms such as conduction through the membrane can be omitted (neglected). Mass transfer of the permeate vapor through the dry membrane micropores is also less subject to resistance since the vacuum applied in the permeate side eliminates most of the trapped air in pores, thus eliminating the molecular diffusion component of the diffusive transport (collision between molecules of gas trapped and permeate vapor), leaving the Knudsen diffusion regime prevailing.

In the AGMD system, the feed side is identical to the DCMD but the permeate side has a stagnant air gap of a certain width followed by a cold surface where permeate vapor condenses. The air gap and associated cold surface are integral parts of the MD module. The cold surface is in fact one side of a cooling system where a coolant continuously recirculates to remove the captured latent heat of the permeate vapor. The existence of an air gap in the permeate side helps reducing conductive heat loss across the membrane but at the same time imparts an additional mass transfer resistance to permeate vapor transport, the magnitude of which depends on the gap width. This concept helps develop the AGMD model equations. It is worth noting that regardless of the configuration or module design, the core model equations are essentially based on simultaneous heat and mass transport phenomena where convective and conductive/diffusive equations are applied where appropriate [4–9].

Finally, in the SGMD system, the feed side is also identical to the DCMD. However, in the permeate side, the permeate vapor is "swept" outside by a sweeping gas and condensed in an external condenser where it is collected. The sweeping gas flowrate and temperature are important operational parameters. The presence of "flowing" sweeping gas in the permeate channel helps reduce the conductive heat loss and minimizes the mass transfer resistance to permeate vapor transport, compared to the concept in the AGMD system. The sweeping gas is often air but can be an inert gas like nitrogen depending on the application. The SGMD is one of the less used configurations, and the few applications reported were mostly for volatile vapors removal from aqueous streams. Limited work was done on desalination using SGMD.

Equipped with the above-mentioned concepts, we can embark on a little more detailed description of membrane distillation model building using examples from extensively used configurations such as the DCMD. However, it is instructive to gain an insight into the various types of models reported in the literature for the benefits of the broader background readership.

8.2 Types of Models for Membrane Distillation

For any system, in general, mathematical models can be classified into two general categories:

- Models based on physical theories. Such models are based on physical-chemical laws involving energy and mass balance, thermodynamics, etc. These are attractive conceptually speaking and can be considered "general" models applicable to any system. However, these models based on pure theory can face difficulty in describing accurately certain complex systems by means of established mathematical relations when the physics in the systems under consideration are poorly understood.
- Models based on purely empirical descriptions, often termed "black box models." These can be devised from simple correlation of input-output, often without physicochemical analysis of the system under

consideration. However, these empirical models, no matter how attractive and convenient they may be, suffer from high risks of extrapolation beyond the range for which they were developed. They certainly cannot be used for scale up directly. Despite this shortfall, empirical models have been of immense help in engineering design.

Certain models can be "hybrids" between models based on physical theory and empirical models. They are hugely popular even if they are not described as such, because they help researchers proceed even when the certain system physics are poorly understood. In these models we see the classic approach of heat and mass conservation approach complemented with empirical correlations of key system parameters like heat and mass transfer coefficients for instance. The empiricism here arises from the uncertain knowledge of the physics governing phenomena within interfaces (e.g., boundary layer between the fluid bulk and membrane surface). This empiricism is also used to "calibrate" models through experimentally determined relationships.

In addition to the broad categories described above, models can be further grouped into five sub-categories:

- Steady state or unsteady state (transient), depending on whether the process investigated has time dependent variables or not.
- Linear or nonlinear. Linear models exhibit superposition while nonlinear models do not. Linear models offer advantages in simplification of mathematics involved.
- Lumped parameter or distributed parameter. Lumped parameter representation is used when spatial variation is non-existent or ignored if justified. This is often exploited in membrane properties that either appear to be nearly uniform or spatial variations are so random that they are ignored for simplicity.
- Continuous or discrete. A continuous variable can assume any value within an interval while a discrete variable can only take distinct values within the interval of interest. Continuous variable modeling prevails in many fields.
- Deterministic or stochastic. A deterministic model predicts a single outcome from a given set of conditions while a stochastic model predicts a set of possible outcomes weighted by their probabilities.

An extra level of categorization could be added: is the model 1 dimensional (1-D) or 2 dimensional (2-D) or three dimensional (3-D)? This is important when physical phenomena variables under study vary along the dimensions. Multiple dimensions make the modeling more complicated and computationally expensive. Unless there are reasons to believe variables and properties vary in multi-dimension systems, the 1-D approach is often chosen through assumptions.

Let us focus on the types of models used in membrane distillation.

Since membrane distillation (MD) is a thermally driven membrane separation process, the backbone of classical mathematical MD models will be based on conservation of mass and energy with simultaneous transport of heat and mass. As yet, no agreed convention has been reported on giving a specific descriptive name to such models. Researchers have so far been free to give generic names of models related to the MD configuration.

One of the earliest coverage of physical phenomena in MD was the paper of Findley et al. published in 1967 [10] who described vaporization through porous media.

Schofield et al. [11] reported on what appears to be among the first heat and mass transfer study in membrane distillation and covering modeling concepts such as type of diffusive transport of vapor through the membrane matrix and the significance of temperature polarization.

Later, a thorough review of membrane distillation covering an introduction to the terminology and fundamental concepts associated with MD was published by Lawson and Lloyd [12]. To date, this is still the most cited review paper on membrane distillation as it covers almost all physical aspects of membrane distillation still discussed nowadays (2018).

Lawson and Lloyd also published some of the earliest detailed papers on direct contact membrane distillation (DCMD) [13] providing experimentally derived membrane coefficients, model predicted flux using pure feed water and vacuum membrane distillation (VMD) [14] also using pure feed water. Using pure water eliminates complications arising from concentration polarization and therefore assists in robust interpretation of observations.

Gryta and Tomaszewska [15] conducted an important heat transport study in MD showing the importance of "validating" heat transfer modeling since much of the heat transfer approach is based on the Nusselt correlations that are numerous and require particular caution in selection. They used a hollow fiber membrane module where the shell side flow is more complex than that of simple flat sheet systems.

Ding et al. [16] conducted a mass transfer study specifically aimed at elucidating the nature of diffusive transport across the membrane in a DCMD system using three different types of membranes and proposed a "new model" for mass transfer in MD. The proposed model was a three-parameter model, named the Knudsen diffusion-molecular diffusion-Poiseuille flow transition (KMPT) model, and was capable to predict the membrane distillation coefficient (MDC) and distillate water flux.

Zhongwei et al. [17] addressed the issue of flow maldistribution in a hollow fiber MD system.

Phattaranawik et al. [9] carried out heat transfer studies in DCMD to evaluate the effect of "spacers" that were used extensively in reverse osmosis to improve mass transfer and diminish concentration polarization and fouling. Their study provided important findings, namely that the membrane distillation coefficient was nearly independent of operational conditions and only depended on the membrane properties. Their model also showed that the temperature distributions were closely linear and their study also concluded that adding the spacer in the flow channel made the flow transitional and not turbulent. In addition, their work seemed to suggest that the membrane effective conductivity was better described by the "alternative" models (also known as isostress and flux law models) rather than the commonly used isostrain model in MD.

Phattaranawik et al. [18,19] carried out DCMD experiments where they observed enhanced flux with spacer used, and pore size distribution enabled a closer study of the type of diffusion regime through the microporous membranes. The model solution showed that Knudsen and transition regions were encountered in the membranes investigated (PVDF, PTFE), while the transition region was the major contribution to mass transfer. Phattaranawik et al. [18] studied the effect of membrane pore size on permeate flux in MD explaining experimental results using the kinetic theory of gases in their modeling approach. Song et al. provided an explanation on how to model hollow fiber DCMD and exploit experimental data obtained [20,21]. Subsequently the review paper of El-Bourawi et al. [22] described what they called "a framework for better understanding membrane distillation separation process." This review paper contained a generous coverage of concepts that were adopted in modeling in numerous papers.

One of the most comprehensive reviews on MD modeling is the paper by Khayet [23] who covered essential membrane properties and how they can be measured and used in MD modeling. The review work covered transport phenomena in the 4 basic MD configurations.

Drioli et al. published a paper on membrane resistance models in DCMD for Knudsen and molecular transport [24]. The approach was hugely popular and adopted in numerous modeling work in MD. Ashoor et al. published a review paper on MD including "black box" and "grey box" modeling [25]. gPROMS, an equation-based modelling and optimization software package developed by Process Systems Enterprise, was used to model MD [26].

However, in a recent review [5], an attempt was made to categorize MD models by virtue of the "approach" adopted in the careful selection of convective heat transfer correlations based on the dimensionless number Nu or Nusselt number in order to extract convective heat transfer coefficients. Numerous correlations have been published [15] and one of the dilemmas would be which of these are better suited to the system under investigation. Some suggestions to alleviate this dilemma have been reported and consist of "validating" the Nu correlation by means of additional experimental work that is purely heat transfer (i.e., using a membrane of the same material as the microporous membrane but is solid, so that the actual material thermal conductivity is used in the overall heat transfer coefficient equation, and not a "corrected effective" coefficient based on assumed theoretical pore models [15]). However, a study of the literature shows that this sensible but slightly costly suggestion has been seldom adopted.

8.3 Model Formulation

Let us illustrate how a classical MD model is formulated for the popular DCMD configuration. This configuration has been the most widely used because of its simplicity in concept and design [23,27]. Extension or modification of the approach can easily be accomplished for other configurations [4].

We previously indicated a simple concept for the DCMD where the system was assumed to be made up of three adjacent parts: the feed channel, the hydrophobic microporous membrane and the permeate channel. We also indicated that MD is a thermally driven process best described by a model concept where simultaneous heat and mass transfer occur from the feed channel to the permeate channel, across the porous membrane. Assuming a "counter-current" flow pattern within the DCMD module, we can conveniently illustrate the three-part MD system graphically as shown in Figure 8.1 and start the model building steps applying simple heat balance with suitable assumptions. In the diagram, we can clearly see that there are process variables that can be directly measured experimentally (feed flowrate, feed temperature in-out; permeate flowrate, permeate temperature in-out), and there are variables that can only be estimated by means of model solution (interfacial temperatures on either side of the membrane). The latter variables are typically located within the module where it is assumed no direct measurements are possible or feasible. Assuming a steady state

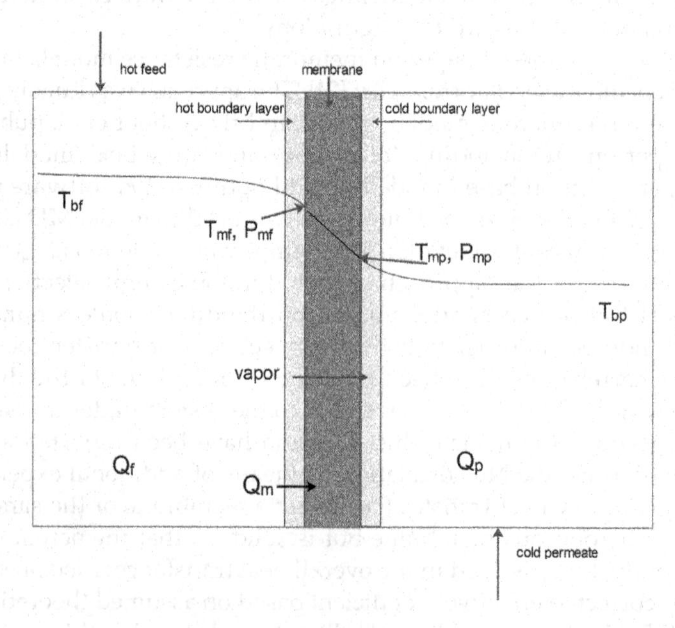

FIGURE 8.1
Mass and heat transport in a counter-current flow DCMD system.

process, heat flows from the hot feed channel to the colder permeate channel through the membrane. The heat flow path is perpendicular to the fluid paths that are assumed counter-current, as shown in Figure 8.1.

The heat lost in the feed channel, Q_f, is equal to the heat transferred through the membrane, Q_m, and also equal to the heat gained in the permeate channel, Q_p. Symbolically:

$$Q_f = Q_m = Q_p \tag{8.1}$$

In order to gain a better insight into the meaning of the three heat flows, we can inspect Figure 8.2 where we clearly see a graphical illustration of the MD module with the terminal temperatures (that can be measured) and a representation of the heat flows Q_f, Q_m and Q_p.

Q_m, the transmembrane heat flow is made up of two components: latent heat of vapor transported Q_{vap} and conductive heat across the membrane Q_{cond}.

Rewriting Q_f, Q_p and Q_m for a flow system:

$$Q_f = m_f C_f (T_{fi} - T_{fo}) \tag{8.2}$$

$$Q_p = m_p C_p (T_{pi} - T_{po}) \tag{8.3}$$

$$Q_m = A_m h_m (T_{fm} - T_{pm}) + A_m J \Delta H_v \tag{8.4}$$

where m_f = Feed mass flow (kg s^{-1}), m_p = Permeate mass flow (kg s^{-1}), C_f = Feed specific heat capacity of the feed (J kg^{-1} K^{-1}), C_p = Feed specific heat capacity of the of the permeate, T_{fi} = Feed inlet temperature (K), T_{fo} = Feed outlet temperature (K), T_{pi} = Permeate inlet temperature (K), T_{po} = Permeate outlet

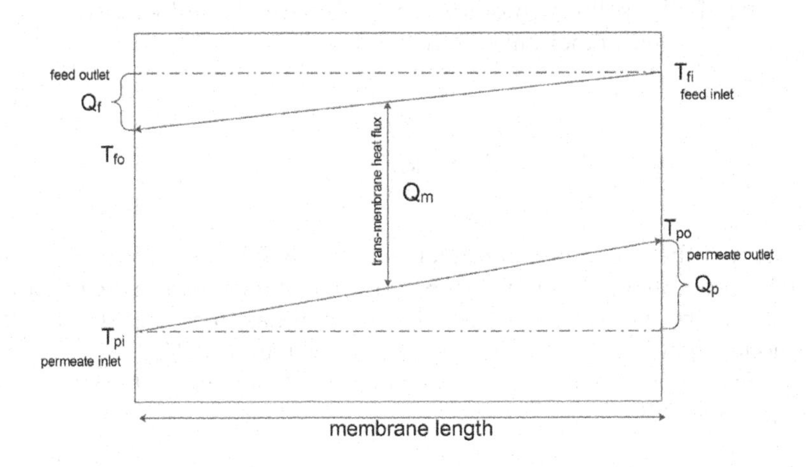

FIGURE 8.2
The three heat flows in a DCMD system.

temperature (K), A_m = Membrane surface area (m^2), h_m = Heat transfer coefficient of the membrane (J s^{-1} m^{-2} K^{-1}), T_{fm} = Feed side membrane surface temperature (K), T_{pm} = Permeate side membrane surface temperature (K), J = Vapor flux (kg m^{-2} s^{-1}) and ΔH_v = Vapor latent heat (J kg^{-1}).

We see that in equation (8.4) there are membrane surface temperatures T_{fm} and T_{pm} that cannot be measured and therefore we need a means to estimate them.

In addition, we notice that transmembrane heat flow term Q_m has unknown quantities that must be determined computationally. The conductive heat transfer coefficient hm is known as the "effective thermal conductivity coefficient" and must be estimated from knowledge of the solid material the membrane is made from and a membrane pore model containing a porosity function. Given that MD membranes can be made using a variety of methods (stretching or phase inversion), it is unlikely that the pore structure is uniform. These pores can be filled by vapor and gas (air) inside the membrane matrix, thus changing its thermal conductivity. Fortunately, the literature has three pore models to choose from [9]:

$$\text{Isostrain model: } h_m = (1-\varepsilon)k_s + \varepsilon k_g \tag{8.5}$$

$$\text{Isostress model: } \left[\frac{\epsilon}{k_g} + \frac{(1-\epsilon)}{k_s}\right]^{-1} \tag{8.6}$$

$$\text{Flux law model: } k_g\left[\frac{1+(1-\epsilon)\beta_{s-g}}{1-(1-\varepsilon)\beta_{s-g}}\right] \tag{8.7}$$

where h_m is the porous membrane thermal conductivity, ε is the membrane porosity, k_s is the solid membrane material thermal conductivity, k_g is the gas (air or otherwise) thermal conductivity and

$$\beta_{s-g} = \frac{k_s/k_g - 1}{k_s/k_g + 2}.$$

It is claimed that the isostrain model (equation 8.5) is the most widely used in MD computations [13,28] but there may be instances where the other two models may be more suitable [9,29]. However, it has to be said that the flux law model equation (8.7) is the least used in the literature [5].

The flux that appears in equation (8.4) can be estimated from equation (8.8) [9,29–33]:

$$j = C_m\left[P_f(T_{fm}, S_{fm}) - (P_p(T_{pm}, S_{pm}))\right] \tag{8.8}$$

where J is the transmembrane permeate flux, C_m is the membrane coefficient (related to a great extent to the membrane properties), S_{fm} and S_{pm} are the salinities on the membrane surface in the feed and permeate sides, P_f is the feed side vapor pressure evaluated at the feed side membrane surface and feed side membrane surface salinity, P_p is the permeate side vapor pressure evaluated at the permeate side membrane surface. For seawater feed, the vapor pressure and many other seawater thermophysical properties can be evaluated using recently published correlations by Nayar et al. [34] within the Lienhard group at MIT [35]. The membrane surface salinity S_{fm} at the feed side can estimated from equation (8.9) [29,33,36]:

$$S_{fm} = S_f exp\left(\frac{J}{\rho_f K}\right) \tag{8.9}$$

where S_f is the feed salinity, J is the permeate flux, ρ_f is the density of the solution and K is the mass transfer coefficient. In the permeate side it is assumed no salt exists as MD theoretically gives 100% salt rejection.

If we wish to gain an insight into the temperature trajectories inside the module to interpret the qualitative temperature curve shown in Figure 8.1, including determination of membrane surface temperatures, we can divide the entire module into N control volumes (slices) of a certain small dimension (typically membrane length divided by N) where physical properties are assumed constant and estimated at an average temperature value between the inlet and outlet of such small control volume (slice), as depicted in Figure 8.3. We can resolve internal temperatures at each control volume through a process of iterative solution of the heat balance equations as shown in the flowchart depicted by Figure 8.4.

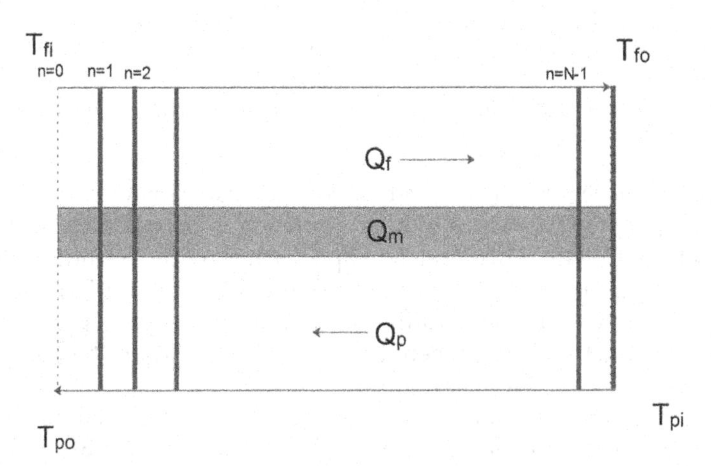

FIGURE 8.3
Division of a membrane into N volume slices for modeling purpose.

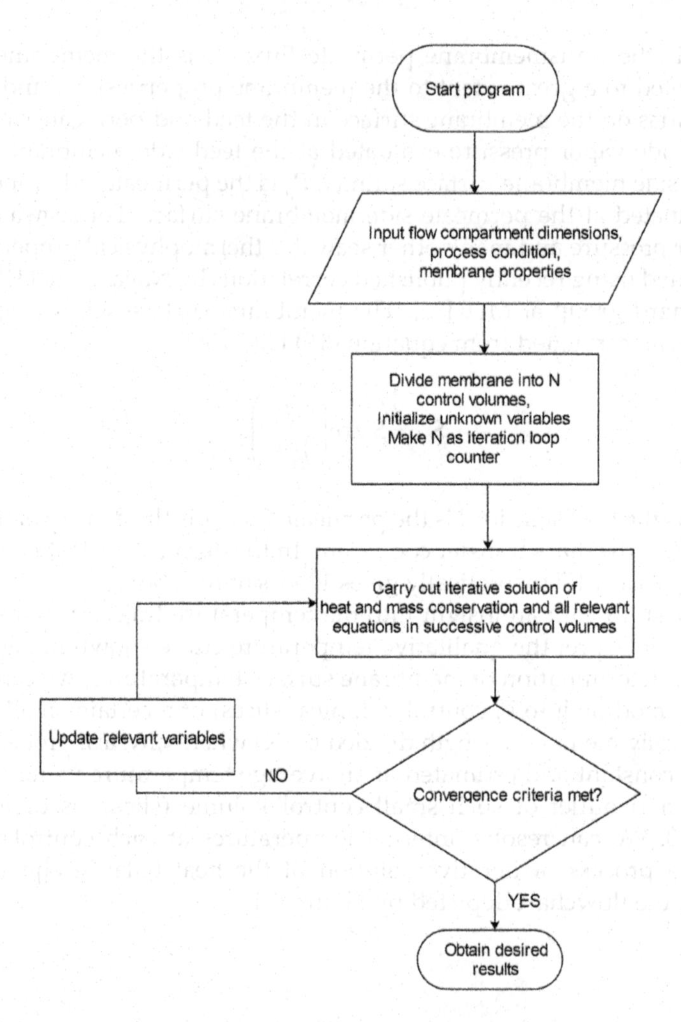

FIGURE 8.4
Flowchart for an of iterative solution of the mass/heat balance equations in membrane distillation modeling.

The complete set of equations and associated correlations can be found in numerous papers [5,29,37]. A similar approach is used for DCMD hollow fiber modules [38–40]. This modeling approach is easily extendable to other module configurations: vacuum membrane distillation (VMD) [41–47], air gap membrane distillation (AGMD) [48–52] and sweeping gas membrane distillation (SGMD) [53,54], where the major differences in the model equations would be in the permeate side.

He et al. reported a simple DCMD model with assumptions that matched experimental data [55]. However, the paper contained little information on the extent of assumptions and certainly the Antoine equation for vapor pressure used does not appear to take into account the feed salinity. For more rigorous estimation of seawater and saline solutions, the paper of Nayar et al. and Sharqawy et al. provide seawater thermophysical properties for modeling and simulation [34,56]. There is even a convenient source for a library of computational routines for the thermophysical properties of seawater within the Lienhard group at MIT [35].

One of the potential uncertainties in MD modeling is the membrane "effective thermal conductivity." Fortunately there has been a number of models reported in the literature in the form of reviews with detailed explanations [5], including how to measure it [57].

Since turbulence is important in MD modeling and can be physically promoted using "spacers" in the MD module, one of earliest and good sources for this topic was covered by Da Costa and Fane [58]. The modeling of turbulence requires access to correlations for heat and mass transfer coefficients, and these were covered comprehensively in a recent paper on MD simulation using CFD [59].

Even though modeling of MD systems incorporates elements of fundamental principles such as conservation of mass and energy, there is a great deal of empirical correlations (for instance to extract thermal and mass transfer coefficients) and a need to input membrane specific parameters. An example of such parameters would be the membrane coefficient, denoted C_m in equation (8.8). C_m, which is specific to the particular type of membrane used, can be determined experimentally from measured flux data or estimated from mass transfer models [5,6,12] where the kinetic theory of gasses [60] was used to determine the membrane coefficient, and subsequently flux performance was determined [12,18,61]. The Dusty Gas model is often used to model the porous structure of hydrophobic membranes [62] in order to determine the membrane coefficient. Thermal conductivities of the "solid" membrane material can easily be found in the literature and sometimes provided by the membrane vendors.

However, much of the modeling work on MD systems in the literature has been carried out in conjunction with accompanying experimental work to validate the model solution, and therefore experimentally determined membrane coefficients appear to be the choice, and quite rightly because the membrane properties (especially the pore structure) vary, sometimes significantly even with the same batch of production. For these reasons, there are in the literature models referred to as experiment "calibrated models" [29]. The literature has also a number of modeling approaches that depart from the traditional approach of solving simultaneous heat and mass transfer equations across the MD membrane that has been divided into control volumes, as described in Figure 8.3. These include Monte Carlo simulation models [63–65], artificial neural network modelling (ANN) [66–68] and CFD modeling of MD [69–86]. Although all of these models might provide the same or at least similar simulation results, the approach in modeling would be different and will probably have certain advantages highlighted in the references provided.

8.4 Models for Various Membrane Distillation Configurations

For the sake of brevity in this work, readers are referred to the abundant literature on specific MD configuration modeling to extract the underlying equations that can be cast in the form of a programming code (Matlab, Fortran, C, etc.) or advanced spreadsheet.

DCMD modeling has been covered in [7,23,29,32,39,40,64,75,77,84,86,87–113], while vacuum MD models were reported in [23,41–47,63,67,69,114–117], including multi-effect [46,47,116]. Modeling on air gap MD can be found in [23,48–52,66,118–121] and modeling on sweeping gas were reported in [23,53,54,71,122,123]. It is worth noting that modeling of MD is becoming popular among graduate students to enhance programming skills and explore MD theory. However, much of the modeling work in the literature is done on clean membranes, thus avoiding complications of membrane fouling modeling that is currently totally absent or approached empirically. There is therefore an opportunity to depart from the tradition of modeling steady state MD processes and engage in the unsteady state nature of MD membrane fouling.

8.5 Models Output

The nature of the model output, also referred to as simulation results, will depend on the objectives set prior to formulating the mathematical model. Hence, the more sophisticated the model, the more equations that describe the physics of the phenomena in the MD module would be required. Most of the time, modeling will be used as a tool to access information that cannot be measured directly, for instance, temperatures on the membrane surface or estimation of the so-called temperature and concentration polarization coefficients. Classical model outputs typically include temperature profiles, concentration profiles and a great deal of intermediate calculation results such as values for Reynolds number, Nusselt number, heat transfer coefficients and mass transfer coefficients, to name a few.

8.6 Main Challenges in Membrane Distillation Modeling

As indicated in the previous section, most of the models are steady state models. However, with the growing interest in solar-powered membrane distillation (SPMD) and the need for fouling membrane modeling, there is a need for dynamic models in MD. There is recent sign of interest in dynamic modeling [101], motivated by cases of intermittent energy supply like solar power. That is indeed a good start. On the other hand, modeling membrane fouling is not straightforward. Much of the model input will consist of yet to be determined heat and mass transfer resistances due to the material build-up on the membrane. As yet, there are no studies that have characterized fully the physical and structural properties of the fouling layer on membranes. This constitutes a golden opportunity to bridge the gap in that respect.

8.7 Emergence of Computational Fluid Dynamics in Membrane Distillation Modeling

Virtually the bulk of MD modeling assumes "average" properties in the computational domains and the dimensions of the computational domain tends to be relatively large, even if the number of control volumes can be a parameter. In fact, some studies showed that a maximum of 10 such control volumes is sufficient and that additional increases would bring any new features to the model results [29]. This is hardly surprising since the flow regime adopted in the flow channels is already "locked" by virtue of selection of the Nusselt number correlation that match the MD thermal performance. This implies that the model will not show any regions that differ in turbulence or mixing even if they exist, due to imperfections in the module design or otherwise. In order to gain an insight into more detailed flow fields in the MD flow channels, a CFD modeling approach has been adopted [69–75,77–80,82–85,86,124–126]. Examples of CFD approaches include new concepts such as multiscale modeling [59], modeling heat and mass transfer in spacer filled channel [74] and studying the internal module geometry [79]. Other studies focused on temperature polarization in MD cells with spacer [127].

8.8 Concluding Remarks

Modeling in MD is the next field of growth alongside membrane development. While steady state modeling is nearly saturated with little new concepts, there are opportunities to engage in dynamic modeling in MD to better understand membrane fouling and better characterize MD performance under unsteady state conditions such as those found in solar-powered membrane distillation or during startup and shutdown of MD units. For a successful start into dynamic modeling of various MD module configurations, there will be a need to input critical data. Such data may include fouling material properties (both physical and structural) and continuous functions of external power input from sources like solar energy. Computational fluid dynamics is a powerful tool that can bring a great deal of benefits in MD modeling and module design. For this tool to make significant advances, there is a need to introduce MD physics in commercially available CFD packages. The learning curve for beginners in CFD modeling can be quite steep and having built-in MD models would lessen the burden of mastering and applying this powerful tool.

References

1. Wang, Q. et al., Desalination by pervaporation: A review. *Desalination*, 2016. **387**: pp. 46–60.
2. Urtiaga, A.M. et al., Parallelism and differences of pervaporation and vacuum membrane distillation in the removal of VOCs from aqueous streams. *Separation and Purification Technology*, 2001. *22–23*: pp. 327–337.
3. Smolders, K. and A.C.M. Franken, Terminology for membrane distillation. *Desalination*, 1989. **72**(3): pp. 249–262.
4. Olatunji, S.O. and L.M. Camacho, Heat and mass transport in modeling membrane distillation configurations: A review. *Frontiers in Energy Research*, 2018. **6**(130).
5. Hitsov, I. et al., Modelling approaches in membrane distillation: A critical review. *Separation and Purification Technology*, 2015. **142**: pp. 48–64.
6. Khayet, M., Membranes and theoretical modeling of membrane distillation: A review. *Advances in Colloid and Interface Science*, 2011. **164**(1–2): pp. 56–88.
7. Qtaishat, M. et al., Heat and mass transfer analysis in direct contact membrane distillation. *Desalination*, 2008. **219**(1–3): pp. 272–292.
8. Curcio, E. and E. Drioli, Membrane distillation and related operations—A review. *Separation and Purification Reviews*, 2005. **34**(1): pp. 35–86.
9. Phattaranawik, J., R. Jiraratananon, and A.G. Fane, Heat transport and membrane distillation coefficients in direct contact membrane distillation. *Journal of Membrane Science*, 2003. **212**(1): pp. 177–193.
10. Findley, M.E., Vaporization through porous membranes. *Industrial & Engineering Chemistry Process Design and Development*, 1967. **6**(2): pp. 226–230.
11. Schofield, R.W., A.G. Fane, and C.J.D. Fell, Heat and mass-transfer in membrane distillation. *Journal of Membrane Science*, 1987. **33**(3): pp. 299–313.
12. Lawson, K.W. and D.R. Lloyd, Membrane distillation. *Journal of Membrane Science*, 1997. **124**(1): pp. 1–25.
13. Lawson, K.W. and D.R. Lloyd, Membrane distillation. II. Direct contact MD. *Journal of Membrane Science*, 1996. **120**(1): pp. 123–133.
14. Lawson, K.W. and D.R. Lloyd, Membrane distillation. I. Module design and performance evaluation using vacuum membrane distillation. *Journal of Membrane Science*, 1996. **120**(1): pp. 111–121.

15. Gryta, M. and M. Tomaszewska, Heat transport in the membrane distillation process. *Journal of Membrane Science*, 1998. **144**(1): pp. 211–222.
16. Ding, Z., R. Ma, and A.G. Fane, A new model for mass transfer in direct contact membrane distillation. *Desalination*, 2003. **151**(3): pp. 217–227.
17. Zhongwei, D., Study on the effect of flow maldistribution on the performance of the hollow fiber modules used in membrane distillation. *Journal of Membrane Science*, 2003. **215**(1–2): pp. 11–23.
18. Phattaranawik, J., Effect of pore size distribution and air flux on mass transport in direct contact membrane distillation. *Journal of Membrane Science*, 2003. **215**(1–2): pp. 75–85.
19. Phattaranawik, J. et al., Mass flux enhancement using spacer filled channels in direct contact membrane distillation. *Journal of Membrane Science*, 2001. **187**(1): pp. 193–201.
20. Song, L. et al., Pilot plant studies of novel membranes and devices for direct contact membrane distillation-based desalination. *Journal of Membrane Science*, 2008. **323**(2): pp. 257–270.
21. Song, L.M. et al., Direct contact membrane distillation-based desalination: Novel membranes, devices, larger-scale studies, and a model. *Industrial & Engineering Chemistry Research*, 2007. **46**(8): pp. 2307–2323.
22. El-Bourawi, M.S. et al., A framework for better understanding membrane distillation separation process. *Journal of Membrane Science*, 2006. **285**(1–2): pp. 4–29.
23. Khayet, M., Membranes and theoretical modeling of membrane distillation: A review. *Advances in Colloid and Interface Science*, 2011. **164**(1–2): pp. 56–88.
24. Drioli, E., A. Ali, and F. Macedonio, Membrane distillation: Recent developments and perspectives. *Desalination*, 2015. **356**: pp. 56–84.
25. Ashoor, B.B. et al., Principles and applications of direct contact membrane distillation (DCMD): A comprehensive review. *Desalination*, 2016. **398**: pp. 222–246.
26. Close, E. and E. Sørensen, Modelling of direct contact membrane distillation for desalination. *20th European Symposium on Computer Aided Process Engineering – ESCAPE20*, 2010. **28**: pp. 649–654.
27. Thomas, N. et al., Membrane distillation research & implementation: Lessons from the past five decades. *Separation and Purification Technology*, 2017. **189**: pp. 108–127.
28. Laganà, F., G. Barbieri, and E. Drioli, Direct contact membrane distillation: Modelling and concentration experiments. *Journal of Membrane Science*, 2000. **166**(1): pp. 1–11.
29. Gustafson, R.D., J.R. Murphy, and A. Achilli, A stepwise model of direct contact membrane distillation for application to large-scale systems: Experimental results and model predictions. *Desalination*, 2016. **378**: pp. 14–27.
30. Martínez-Díez, L., M.I. Vázquez-González, and F.J. Florido-Díaz, Temperature polarization coefficients in membrane distillation. *Separation Science and Technology*, 1998. **33**(6): pp. 787–799.
31. Burgoyne, A. and M.M. Vahdati, Direct contact membrane distillation. *Separation Science and Technology*, 2000. **35**(8): pp. 1257–1284.
32. Bui, V.A., L.T.T. Vu, and M.H. Nguyen, Simulation and optimisation of direct contact membrane distillation for energy efficiency. *Desalination*, 2010. **259**(1): pp. 29–37.

33. Suárez, F., S.W. Tyler, and A.E. Childress, A theoretical study of a direct contact membrane distillation system coupled to a salt-gradient solar pond for terminal lakes reclamation. *Water Research*, 2010. **44**(15): pp. 4601–4615.
34. Nayar, K.G. et al., Thermophysical properties of seawater: A review and new correlations that include pressure dependence. *Desalination*, 2016. **390**: pp. 1–24.
35. Lienhard V, J.H., Thermophysical properties of seawater. 2017 [citedOctober 13, 2018]. Available from: http://web.mit.edu/seawater/.
36. Yun, Y. et al., Direct contact membrane distillation mechanism for high concentration NaCl solutions. *Desalination*, 2006. **188**(1–3): pp. 251–262.
37. Suarez, F., S.W. Tyler, and A.E. Childress, A theoretical study of a direct contact membrane distillation system coupled to a salt-gradient solar pond for terminal lakes reclamation. *Water Research*, 2010. **44**(15): pp. 4601–4615.
38. Cheng, L.H., P.C. Wu, and J.H. Chen, Modeling and optimization of hollow fiber DCMD module for desalination. *Journal of Membrane Science*, 2008. **318**(1–2): pp. 154–166.
39. Ibrahim, S.S. and Q.F. Alsalhy, Modeling and simulation for direct contact membrane distillation in hollow fiber modules. *AIChE Journal*, 2013. **59**(2): pp. 589–603.
40. Bui, V.A., L.T.T. Vu, and M.H. Nguyen, Modelling the simultaneous heat and mass transfer of direct contact membrane distillation in hollow fibre modules. *Journal of Membrane Science*, 2010. **353**(1–2): pp. 85–93.
41. Bandini, S., A. Saavedra, and G.C. Sarti, Vacuum membrane distillation: Experiments and modeling. *Aiche Journal*, 1997. **43**(2): pp. 398–408.
42. Qi, B., B. Li, and S. Wang, Investigation of shell side heat transfer in cross-flow designed vacuum membrane distillation module. *Industrial & Engineering Chemistry Research*, 2012. **51**(35): pp. 11463–11472.
43. Criscuoli, A., M.C. Carnevale, and E. Drioli, Modeling the performance of flat and capillary membrane modules in vacuum membrane distillation. *Journal of Membrane Science*, 2013. **447**: pp. 369–375.
44. Lee, J.-G. and W.-S. Kim, Numerical modeling of the vacuum membrane distillation process. *Desalination*, 2013. **331**: pp. 46–55.
45. Kim, Y.-D., Y.-B. Kim, and S.-Y. Woo, Detailed modeling and simulation of an out-in configuration vacuum membrane distillation process. *Water Research*, 2018. **132**: pp. 23–33.
46. Kiefer, F., M. Spinnler, and T. Sattelmayer, Multi-effect vacuum membrane distillation systems: Model derivation and calibration. *Desalination*, 2018. **438**: pp. 97–111.
47. Boutikos, P. et al., A theoretical approach of a vacuum multi-effect membrane distillation system. *Desalination*, 2017. **422**: pp. 25–41.
48. He, Q.F. et al., Modeling and optimization of air gap membrane distillation system for desalination. *Desalination*, 2014. **354**: pp. 68–75.
49. Dehesa-Carrasco, U., C.A. Perez-Rabago, and C.A. Arancibia-Bulnes, experimental evaluation and modeling of internal temperatures in an air gap membrane distillation unit. *Desalination*, 2013. **326**: pp. 47–54.
50. Khayet, M. and C. Cojocaru, Air gap membrane distillation: Desalination, modeling and optimization. *Desalination*, 2012. **287**: pp. 138–145.
51. Chang, H. et al., Experimental and simulation study of an air gap membrane distillation module with solar absorption function for desalination. *Desalination and Water Treatment*, 2011. **25**(1–3): pp. 251–258.

52. Guijt, C. et al., Air gap membrane distillation 2. Model validation and hollow fibre module performance analysis. *Separation and Purification Technology*, 2005. **43**(3): pp. 245–255.
53. Karanikola, V. et al., Sweeping gas membrane distillation: Numerical simulation of mass and heat transfer in a hollow fiber membrane module. *Journal of Membrane Science*, 2015. **483**: pp. 15–24.
54. Khayet, M., C. Cojocaru, and A. Baroudi, Modeling and optimization of sweeping gas membrane distillation. *Desalination*, 2012. **287**: pp. 159–166.
55. He, K. et al., Production of drinking water from saline water by direct contact membrane distillation (DCMD). *Journal of Industrial and Engineering Chemistry*, 2011. **17**(1): pp. 41–48.
56. Sharqawy, M.H., J.H. Lienhard, and S.M. Zubair, Thermophysical properties of seawater: A review of existing correlations and data. *Desalination and Water Treatment*, 2010. **16**(1–3): pp. 354–380.
57. García-Payo, M.C. and M.A. Izquierdo-Gil, Thermal resistance technique for measuring the thermal conductivity of thin microporous membranes. *Journal of Physics D: Applied Physics*, 2004. **37**(21): p. 3008.
58. Da Costa, A.R., A.G. Fane, and D.E. Wiley, Spacer characterization and pressure drop modelling in spacer-filled channels for ultrafiltration. *Journal of Membrane Science*, 1994. **87**(1): pp. 79–98.
59. Amigo, J., R. Urtubia, and F. Suárez, Exploring the interactions between hydrodynamics and fouling in membrane distillation systems—A multiscale approach using CFD. *Desalination*, 2018. **444**: pp. 63–74.
60. Present, R.D., *Kinetic Theory of Gases*. 1958, New York: McGraw-Hill.
61. Khayet, M., A. Velázquez, and J.I. Mengual, Modelling mass transport through a porous partition: Effect of pore size distribution. *Journal of Non-Equilibrium Thermodynamics*, 2004. **29**(3).
62. Mason, E.A. and M.A. Malinauskas, Gas Transport in Porous Media: The Dusty-gas Model. Chemical Engineering Monographs . 1983, Elsevier.
63. Imdakm, A.O., M. Khayet, and T. Matsuura, A monte carlo simulation model for vacuum membrane distillation process. *Journal of Membrane Science*, 2007. **306**(1–2): pp. 341–348.
64. Imdakm, A., A Monte Carlo simulation model for membrane distillation processes: Direct contact (MD). *Journal of Membrane Science*, 2004. **237**(1–2): pp. 51–59.
65. Khayet, M., A.O. Imdakm, and T. Matsuura, Monte carlo simulation and experimental heat and mass transfer in direct contact membrane distillation. *International Journal of Heat and Mass Transfer*, 2010. **53**(7–8): pp. 1249–1259.
66. Khayet, M. and C. Cojocaru, Artificial neural network modeling and optimization of desalination by air gap membrane distillation. *Separation and Purification Technology*, 2012. **86**: pp. 171–182.
67. Cao, W.S. et al., Modeling and simulation of VMD desalination process by ANN. *Computers & Chemical Engineering*, 2016. **84**: pp. 96–103.
68. Khayet, M. and C. Cojocaru, Artificial neural network model for desalination by sweeping gas membrane distillation. *Desalination*, 2013. **308**: pp. 102–110.
69. Hayer, H., O. Bakhtiari, and T. Mohammadi, Analysis of heat and mass transfer in vacuum membrane distillation for water desalination using computational fluid dynamics (CFD). *Desalination and Water Treatment*, 2015. **55**(1): pp. 39–52.
70. Chang, H. et al., CFD study of heat transfer enhanced membrane distillation using spacer-filled channels. *Energy Procedia*, 2015. **75**: pp. 3213–3219.

71. Jafari, P. and M. Keshavarz Moraveji, Application of generic cubic equations of state in the CFD simulation of the sweeping gas polytetrafluoroethylene (PTFE) membrane distillation. *Desalination and Water Treatment*, 2014. **57**(4): pp. 1647–1658.
72. Gurreri, L. et al., CFD prediction of concentration polarization phenomena in spacer-filled channels for reverse electrodialysis. *Journal of Membrane Science*, 2014. **468**: pp. 133–148.
73. Shakaib, M. et al., A CFD study of heat transfer through spacer channels of membrane distillation modules. *Desalination and Water Treatment*, 2013. **51**(16–18): pp. 3662–3674.
74. Al-Sharif, S. et al., Modelling flow and heat transfer in spacer-filled membrane distillation channels using open source CFD code. *Desalination*, 2013. **311**: pp. 103–112.
75. Yu, H. et al., Analysis of heat and mass transfer by CFD for performance enhancement in direct contact membrane distillation. *Journal of Membrane Science*, 2012. **405**: pp. 38–47.
76. Yang, X. et al., Analysis of the effect of turbulence promoters in hollow fiber membrane distillation modules by computational fluid dynamic (CFD) simulations. *Journal of Membrane Science*, 2012. *415–416*: pp. 758–769.
77. Yang, X. et al., Optimization of microstructured hollow fiber design for membrane distillation applications using CFD modeling. *Journal of Membrane Science*, 2012. *421–422*: pp. 258–270.
78. Shakaib, M. et al., A CFD study on the effect of spacer orientation on temperature polarization in membrane distillation modules. *Desalination*, 2012. **284**: pp. 332–340.
79. Cipollina, A., G. Micale, and L. Rizzuti, Membrane distillation heat transfer enhancement by CFD analysis of internal module geometry. *Desalination and Water Treatment*, 2011. **25**(1–3): pp. 195–209.
80. Cipollina, A. et al., CFD simulation of a membrane distillation module channel. *Desalination and Water Treatment*, 2009. **6**(1–3): pp. 177–183.
81. Katsandri, A., A theoretical analysis of a spacer filled flat plate membrane distillation modules using CFD: Part II: Temperature polarisation analysis. *Desalination*, 2017. **408**: pp. 166–180.
82. Shirazi, M.M.A. et al., Computational Fluid Dynamic (CFD) opportunities applied to the membrane distillation process: State-of-the-art and perspectives. *Desalination*, 2016. **377**: pp. 73–90.
83. Jafari, P. and M.K. Moraveji, Application of generic cubic equations of state in the CFD simulation of the sweeping gas polytetrafluoroethylene (PTFE) membrane distillation. *Desalination and Water Treatment*, 2016. **57**(4): pp. 1647–1658.
84. Chang, H., C.-D. Ho, and J.-A. Hsu, Analysis of heat transfer coefficients in direct contact membrane distillation modules using CFD simulation. *Journal of Applied Science and Engineering*, 2016. **19**(2): pp. 197–206.
85. Yazgan-Birgi, P., M.I. Hassan Ali, and H.A. Arafat, Estimation of liquid entry pressure in hydrophobic membranes using CFD tools. *Journal of Membrane Science*, 2018. **552**: pp. 68–76.
86. Rezakazemi, M., CFD simulation of seawater purification using direct contact membrane desalination (DCMD) system. *Desalination*, 2018. **443**: pp. 323–332.

87. Soukane, S., J.-G. Lee, and N. Ghaffour, Direct contact membrane distillation module scale-up calculations: Choosing between convective and conjugate approaches. *Separation and Purification Technology*, 2019. **209**: pp. 279–292.
88. Soomro, M.I. and W.-S. Kim, Performance and economic evaluation of linear Fresnel reflector plant integrated direct contact membrane distillation system. *Renewable Energy*, 2018. **129**: pp. 561–569.
89. Long, R. et al., Direct contact membrane distillation system for waste heat recovery: Modelling and multi-objective optimization. *Energy*, 2018. **148**: pp. 1060–1068.
90. Lee, J.-G. et al., Influence of high range of mass transfer coefficient and convection heat transfer on direct contact membrane distillation performance. *Desalination*, 2018. **426**: pp. 127–134.
91. Taamneh, Y. and K. Bataineh, Improving the performance of direct contact membrane distillation utilizing spacer-filled channel. *Desalination*, 2017. **408**: pp. 25–35.
92. Santoro, S. et al., A non-invasive optical method for mapping temperature polarization in direct contact membrane distillation. *Journal of Membrane Science*, 2017. **536**: pp. 156–166.
93. Deshpande, J., K. Nithyanandam, and R. Pitchumani, Analysis and design of direct contact membrane distillation. *Journal of Membrane Science*, 2017. **523**: pp. 301–316.
94. Imdakm, A.O. and T. Matsuura, Corrigendum to "Simulation of heat and mass transfer in direct contact membrane distillation (MD): The effect of membrane physical properties." *Journal of Membrane Science*, 2006. **270**(1–2): pp. 228.
95. Imdakm, A.O. and T. Matsuura, Simulation of heat and mass transfer in direct contact membrane distillation (MD): The effect of membrane physical properties (vol 262, pg 117, 2005). *Journal of Membrane Science*, 2006. **270**(1–2): pp. 228–228.
96. Imdakm, A.O. and T. Matsuura, Simulation of heat and mass transfer in direct contact membrane distillation (MD): The effect of membrane physical properties. *Journal of Membrane Science*, 2005. **262**(1): pp. 117–128.
97. Burgoyne, A. and M.M. Vahdati, Permeate flux modeling of membrane distillation. *Filtration & Separation*, 1999. **36**(1): pp. 49–53.
98. Banat, F.A. et al., Modeling of desalination using tubular direct contact membrane distillation modules. *Separation Science and Technology*, 1999. **34**(11): pp. 2191–2206.
99. Lee, J.-G. et al., Dynamic solar-powered multi-stage direct contact membrane distillation system: Concept design, modeling and simulation. *Desalination*, 2018. **435**: pp. 278–292.
100. Karam, A.M. and T.M. Laleg-Kirati, Electrical equivalent thermal network for direct contact membrane distillation modeling and analysis. *Journal of Process Control*, 2016. **47**: pp. 87–97.
101. Eleiwi, F. et al., Dynamic modeling and experimental validation for direct contact membrane distillation (DCMD) process. *Desalination*, 2016. **384**: pp. 1–11.
102. Cheng, D., W. Gong, and N. Li, Response surface modeling and optimization of direct contact membrane distillation for water desalination. *Desalination*, 2016. **394**: pp. 108–122.

103. Lee, J.-G. et al., Performance modeling of direct contact membrane distillation (DCMD) seawater desalination process using a commercial composite membrane. *Journal of Membrane Science*, 2015. **478**: pp. 85–95.
104. Lin, S., N.Y. Yip, and M. Elimelech, Direct contact membrane distillation with heat recovery: Thermodynamic insights from module scale modeling. *Journal of Membrane Science*, 2014. **453**: pp. 498–515.
105. Ghadiri, M., S. Fakhri, and S. Shirazian, Modeling of water transport through nanopores of membranes in direct-contact membrane distillation process. *Polymer Engineering & Science*, 2014. **54**(3): pp. 660–666.
106. Jeong, S. et al., Structural analysis and modeling of the commercial high performance composite flat sheet membranes for membrane distillation application. *Desalination*, 2014. **349**: pp. 115–125.
107. Boubakri, A., A. Hafiane, and S.A. Bouguecha, Application of response surface methodology for modeling and optimization of membrane distillation desalination process. *Journal of Industrial and Engineering Chemistry*, 2014. **20**(5): pp. 3163–3169.
108. Kurdian, A.R. et al., Modeling of direct contact membrane distillation process: Flux prediction of sodium sulfate and sodium chloride solutions. *Desalination*, 2013. **323**: pp. 75–82.
109. Ho, C.D., T.J. Yang, and B.C. Wang, Modeling of conjugated heat transfer in direct-contact membrane distillation of seawater desalination systems. *Chemical Engineering & Technology*, 2012. **35**(10): pp. 1765–1776.
110. Hwang, H.J. et al., Direct contact membrane distillation (DCMD): Experimental study on the commercial PTFE membrane and modeling. *Journal of Membrane Science*, 2011. **371**(1–2): pp. 90–98.
111. Chen, T.-C., C.-D. Ho, and H.-M. Yeh, Theoretical modeling and experimental analysis of direct contact membrane distillation. *Journal of Membrane Science*, 2009. **330**(1–2): pp. 279–287.
112. Cheng, L.-H., P.-C. Wu, and J. Chen, Modeling and optimization of hollow fiber DCMD module for desalination. *Journal of Membrane Science*, 2008. **318**(1–2): pp. 154–166.
113. Bandini, S., C. Gostoli, and G.C. Sarti, Role of heat and mass-transfer in membrane distillation process. *Desalination*, 1991. **81**(1–3): pp. 91–106.
114. Ma, Q., A. Ahmadi, and C. Cabassud, Direct integration of a vacuum membrane distillation module within a solar collector for small-scale units adapted to seawater desalination in remote places: Design, modeling & evaluation of a flat-plate equipment. *Journal of Membrane Science*, 2018. **564**: pp. 617–633.
115. Cheng, D., N. Li, and J. Zhang, Modeling and multi-objective optimization of vacuum membrane distillation for enhancement of water productivity and thermal efficiency in desalination. *Chemical Engineering Research and Design*, 2018. **132**: pp. 697–713.
116. Zhang, Y. et al., Numerical modeling and economic evaluation of two multi-effect vacuum membrane distillation (ME-VMD) processes. *Desalination*, 2017. **419**: pp. 39–48.
117. Zuo, G., G. Guan, and R. Wang, Numerical modeling and optimization of vacuum membrane distillation module for low-cost water production. *Desalination*, 2014. **339**: pp. 1–9.
118. Bindels, M., N. Brand, and B. Nelemans, Modeling of semibatch air gap membrane distillation. Desalination, 2018. **430**: pp. 98–106.

119. Cheng, L.H., Y.H. Lin, and J.H. Chen, Enhanced air gap membrane desalination by novel finned tubular membrane modules. *Journal of Membrane Science*, 2011. **378**(1–2): pp. 398–406.
120. Alsaadi, A.S. et al., Modeling of air-gap membrane distillation process: A theoretical and experimental study. *Journal of Membrane Science*, 2013. **445**: pp. 53–65.
121. Chang, H.A. et al., Modeling and optimization of a solar driven membrane distillation desalination system. *Renewable Energy*, 2010. **35**(12): pp. 2714–2722.
122. Perfilov, V., V. Fila, and J. Sanchez Marcano, A general predictive model for sweeping gas membrane distillation. *Desalination*, 2018. **443**: pp. 285–306.
123. Karanikola, V. et al., Effects of membrane structure and operational variables on membrane distillation performance. *Journal of Membrane Science*, 2017. **524**: pp. 87–96.
124. Katsandri, A., A theoretical analysis of a spacer filled flat plate membrane distillation modules using CFD: Part I: velocity and shear stress analysis. *Desalination*, 2017. **408**: pp. 145–165.
125. Gurreri, L. et al., Flow and mass transfer in spacer-filled channels for reverse electrodialysis: A CFD parametrical study. *Journal of Membrane Science*, 2016. **497**: pp. 300–317.
126. Tamburini, A. et al., CFD prediction of scalar transport in thin channels for reverse electrodialysis. *Desalination and Water Treatment*, 2015. **55**(12): pp. 3424–3445.
127. Tamburini, A. et al., A thermochromic liquid crystals image analysis technique to investigate temperature polarization in spacer-filled channels for membrane distillation. *Journal of Membrane Science*, 2013. **447**: pp. 260–273.

9

Low-Carbon Energy Sources for Membrane Distillation Processes for Desalination

9.1 Introduction

In Chapter 1 we saw that desalination is energy intensive and has a practical lower limit on energy consumption of 1.56 kWh/m^3 (RO with 50% recovery) [1] that is hard to beat no matter what desalination technology is employed. Reverse osmosis, with the best energy exchangers available to date and its "gold standard" thin composite film semi permeable membranes can manage an energy consumption of 2 kWh/m^3 [1]. In addition, the best RO plants with an energy consumption ranging from 2 to 4 kWh/m^3 emit carbon dioxide in the range of 0.92–1.78 kg CO_2/m^3 [2]. This emission, however undesirable it may be, is considerably lower than for the conventional thermal desalination technologies MSF and MED that emit 23.41 and 18.05 kg CO_2/m^3 respectively [2]. The carbon footprint is the second concern after cost (capital and operational) in conventional desalination technologies (thermal and reverse osmosis).

So where does membrane distillation stand with respect to cost and carbon footprint? At the present it hard to provide a firm answer since MD is not deployed commercially and any claims or published data on this matter would be at best speculative or indicative. However, one can engage in the topic through indirect channels to study the relative merits of MD with respect to operational cost and carbon footprint. At this point it is worth recalling that MD has been earmarked right from its early inception as a "niche market" for remote locations and for specialist applications rather than a direct competitor to established desalination technologies such as reverse osmosis [3]. The main advantages as well as the outstanding negative issues that still need addressing in membrane distillation for desalination have been highlighted in the previous chapters. In this chapter, we will focus on those sources of energy for membrane distillation for desalination that can be considered "game changers," in particular at a time when low carbon technologies are well sought after to mitigate the carbon conundrum [4].

We will start by paying particular attention to a form of heat energy that is currently dissipated into the environment as "low-grade waste heat" because the stream temperatures are not considered suitable for the "parent" process, and thereafter, we refocus our attention to solar energy that is considered the most sustainable form of energy (clean and abundant).

9.1.1 Low-Grade Waste Heat

Forman et al. [5] reported that the process "chain of energy conversion" from primary energy carriers to final energy use is subject to a number of losses. In particular, in end use, significant amounts of converted energy end up as waste heat released to the environment because it is unwanted. In order to improve energy efficiency and reduce energy consumption, the so-called waste heat has to be utilized in some way. Forman et al. [5] estimated that 72% of the global primary energy consumption is lost after conversion. In addition, 63% of the waste heat streams considered in their studies arise at a temperature below 100°C. In another study on low-grade waste heat in the United Kingdom alone, Law et al. [6] found that significant gains can be made in this sector by recovering low-grade waste heat, as up to 14 TWh per annum, representing 4% of total energy use, of the UK process industries' energy consumption is currently lost as recoverable waste heat. Rattner reported in his Penn State University research web site that the United States has a potential for 13×10^9 Gj of waste heat recovery of which 10.3×10^9 Gj accounts for low-grade heat below 100°C and above 40°C [7]. These results are absolutely staggering and demonstrate not only the massive scale of heat losses to the environment from process industries, commercial and residential space heating and air conditioning, but also the huge potential and opportunities to tap into this underused energy source. The idea of exploiting low-grade waste heat for conventional desalination has been reported in the literature for a number of years [8–10]. However, there has not been an attempt to implement any of the proposed schemes for a variety of reasons that include lack of budgeting cost and lack of comprehensive investment risk assessment.

9.1.2 Solar Energy Harvesting for Desalination

One of the earliest reviews on solar desalination is that of Talbert et al. [11] who provided a useful manual of solar distillation. That was in 1970. Since then, a great deal has been published on solar distillation. The review by Tiwari et al. [12] on the status of solar distillation provides an update on the solar technologies and the basic heat and mass transfer relation responsible for developing and testing procedure for various designs of solar stills. More recent reviews on solar stills/collectors and solar desalination demonstrate the surge in interest in this clean and sustainable form of energy to treat water and produce clean freshwater [13–21]. Commercial solar thermal collectors are now widely available and prices have come down considerably due to the intense competition in the market. They are also available in a wide range of configurations (flat plate, evacuated tube, etc.). At this point it is important to recall that for solar energy research work, reliable and easy to find sources of solar irradiance data are extremely important in the early stages of the work to conduct exploratory simulations. An excellent, convenient and free to use (requires registration) source is the "The Renewables. ninja" [22].

9.1.3 Low-Grade Waste and Solar Energy Recovery for Membrane Distillation Desalination

While solar energy is freely available anywhere and low-grade waste heat is also freely available within the process it is released from, both actually require a heat recovery infrastructure to make it available to any utilization, including for membrane distillation desalination. The simplified diagram shown in Figure 9.1. There are, however, some important differences between the two heat recovery systems: the industrial low-grade waste heat tends to be continuous, stable and predictable while solar energy is dynamic, discontinuous and prone to climatic and environmental disruptions, such as seasonal solar irradiance variation, clouds, dust, variable wind speeds, etc. For these reasons, the literature is in need for solid modeling and simulation of thermal integration of membrane distillation to a variety of low-grade

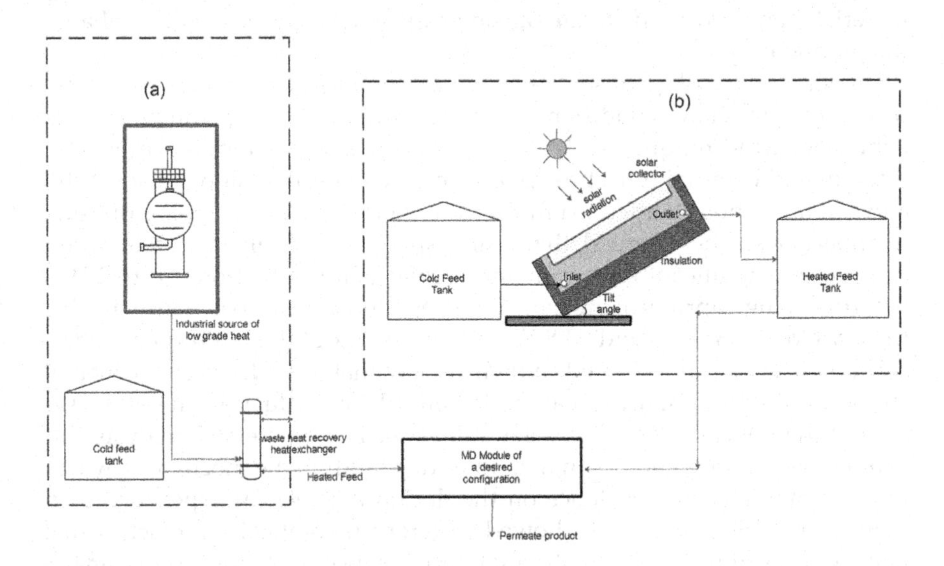

FIGURE 9.1
Thermally integrated membrane distillation module with a process plant releasing low-grade waste heat (a) and a solar collector system heating the feed to a membrane distillation module (b).

waste heat sources and to solar collector systems under a variety of location, climatic and environmental conditions to predict performance before any hard investment is made.

Pugsley et al. wrote a chapter on solar potential around the world [13] and provided valuable data on water scarcity, solar insolation and an interesting applicability index of solar desalination globally. Interestingly, the applicability score (index) nearly matched water stressed parts of the world.

Kabeel and Al-Agouz reported on a single-effect solar still, that is a simple and low cost solar device used for converting the available brackish or waste water into potable water [21]. They also reported on ways to enhance productivity of such simple solar distillation devices.

Prakash and Velmurugan [18] published a review on the various parameters influencing the productivity of the solar stills. They described factors for improving productivity such as area of absorption, minimum depth of water, water–glass cover temperature difference, inlet water temperature, heat storage, phase change materials, vacuum technology and other methods such as using reflectors, condensers and multi effect distillation. Their analysis showed that the productivity increases when the area of the absorption is increased. The basin water depth was found to be the main parameter that affects the productivity of the still. The increase of the water–glass cover temperature difference ($Tw-Tg$) was also found to play a significant role in increasing the productivity. Preheating the feed water to the still basin showed a considerable improvement in the productivity. The literature studied showed that solar stills with heat storage medium and phase change materials can produce distillate outside sunshine hours and hence, enhance the productivity.

Sharshir et al. [23] presented a review of factors affecting solar still production (climatic conditions, operations and design parameters) and enhancement techniques such as wicks, internal and external condensers, internal and external reflectors, phase change materials, stepped solar still and a new method claimed to improve the solar still yield by means of using nanoparticles. They argued that using sponge cubes in the basin water can cause a significant enhancement in solar still production (up to 273%) whereas using cuprous oxide nanoparticles was claimed to increase the distilled yield by 133.64% and 93.87% with and without the fan respectively.

Sharshir et al. [15] indicated in their review that apart from experimental studies, theoretical analysis can be beneficial for evaluating the effectiveness of theoretically designed solar stills (SS). The theoretical analysis and simulation modeling of SS can help in understanding the mechanics, which in turn, may provide guidance on the design of practical experiments. In their review, Sharshir et al. showed different theoretical approaches that have been used to assess the thermal performance of SS and exergy analysis of such devices were discussed. The review work indicated that the productivity of the SS depended on different external and internal operating parameters. They pointed out that the distillate quantity varied according to certain design features and also to related technical advancements of the SS. Sharshir et al. [15] stated that the objective of the review paper was to highlight design methods that could ultimately allow the researchers to optimize the SS for further development.

Qtaishat and Banat [24] wrote a review article in which they reported membrane distillation fundamentals coupled with literature on solar energy heating of the feed using well known examples of systems that use solar collectors and Khayet [25] wrote a detailed review article showing the scatter in data on cost of production of water using MD and energy consumption. The paper certainly pointed at some weaknesses in the literature where result reporting in MD work did not follow a systematic procedure to provide all relevant details to enable readers to compare like with like. That is indeed unfortunate.

Exergy analysis in MD is increasingly studied and Gude [26] published a recent paper that elaborates on use of exergy tools to evaluate renewable energy powered desalination processes to evaluate their thermodynamic efficiency. Illustrations were provided to identify the major components and process streams that contributed to major exergy destruction and to ultimately suggest suitable operating conditions that minimize exergy losses. Traditional methods such as MSF, MED, RO, solar distillation as well as the emerging membrane distillation technologies were discussed with case studies to illustrate their exergy performances and hopefully optimize them. The analysis of Gude showed remarkably well the benefits of integrating membrane processes with power generation or legacy thermal desalination processes. The work also showed that the exergy losses are the highest for the solar energy-based systems, because the solar exergy is quite high.

9.2 Low-Grade Heat Sources and Utilization in Membrane Distillation

Despite acknowledging the importance and availability of low-grade waste heat for membrane distillation, very little work has been done and published on this topic compared to the vast growth of publications on work done in the laboratory using electrical heating of the feed to simulate feed temperatures that would be encountered if real waste heat sources were to be employed. In this section we will explore some of the recently published work on low-grade waste potential or actual use in membrane distillation for desalination. The literature to be covered in this section will include some work on the so-called niche applications such as desalination onboard ships where prospects should be excellent.

Xu et al. [27] published a paper describing a pilot plant scale vacuum membrane distillation (VMD) system, using polypropylene hollow fiber.

The device was designed and installed on a ship to investigate the operating conditions. During operation, the feed seawater was heated by the waste heat generated from the ship engine. The authors reported that this VMD system could achieve a salt rejection of 99.99% and a membrane flux of 5.4 kg/m^2.h at 55°C feed temperature and a permeate side vacuum of −0.093 Mpa. The results obtained from VMD pilot test are in accord with the previous lab experiment results. This work was one rare instance of membrane distillation desalination testing onboard a real ship.

Amaya-Vías et al. [28] carried out an experimental study involving three MD configurations (water gap MD, direct contact MD and air gap MD) using 2 PTFE membranes (0.2 and 0.45 μ pore diameter) and applying conditions that simulated conditions onboard cruise ships with an objective to affirm the suitability of these MD modules for use onboard ships to produce freshwater exploiting engine waste heat. They argue that the cruise tourism industry is expanding worldwide and that an average cruise ship produces over 1000 m^3 per day of freshwater. They stated that to date, multi-stage flash distillation (MSF) and seawater reverse osmosis (SWRO) were the only desalination technologies on board ships. Their results were in line with the literature on MD, showing that the flux and salt rejection were reasonable in all cases studied (feed/permeate side temperatures of 75°C/18°C) and similar to results published in the literature for similar devices and membrane material and properties. They concluded that membrane distillation has a good potential onboard ships to produce freshwater. However, what was lacking in their work was the long-term performance and how to deal with membrane fouling/scaling.

Gonzalez et al. carried out a critical review on the application of MD involving low-grade heat (and renewable energy sources) [29]. Their article focused on applications for sustainable water production and on issues that must be addressed to improve the performance of MD desalination systems and possibly achieve zero liquid discharge. They identified well known research gaps in membrane distillation desalination, namely, new membranes, enhanced membrane modules and novel MD configurations. This work is a useful confirmation of previously published material on the use (or potential use) of renewable energy sources and low-grade heat in MD.

Long et al. [30] published a recent paper on the potential of low-grade heat in relation to MD performance optimization. The evaluation of performance of a DCMD system with heat recovery, gain output ration (GOR) and mass recovery rate are normally the two main criteria. However, they could not achieve their maximum values simultaneously. To achieve a compromise, a multi-objective optimization study considering both the water recovery rate and GOR was carried out. In addition to GOR, mass recovery rate and thermal efficiency under single-objective optimization methods were computed and compared. In comparison to the results obtained under maximum GOR, the increase in water mass recovery rate under the multi-objective optimization compensated the decrease magnitude of GOR.

Compared with the performance under the single-objective optimization for transmembrane water flux, the transmembrane water flux under the multi-objective optimization was reduced by only 6.7%, but the GOR is increased by 83.2%. This paper is one of several theoretical investigations under the topic of waste heat utilization.

Lai et al. [31] proposed a novel concept where a hybrid system consisting of PEMFC (proton exchange membrane fuel cell) and DCMD (direct contact membrane distillation) were combined to recover the waste heat from PEMFC for brine water desalination. They investigated parameters that determine the performance of this hybrid system and found that an optimum PEMFC current density and DCMD fresh water inlet mass flow rate can exist thus leading to a maximum energy gain from the fuel chemical energy. Lai et al. [31] used a genetic algorithm method in order to analyze the optimal performance of the proposed hybrid system to obtain the optimal PEMFC current density and DCMD fresh solution inlet mass flow rate. They observed that the energy utilization degree can be increased by 201%–266% when the operating temperature varies from 328.15 K to 348.15 K. This is yet another example of the use of simulation to explore the best way to thermally integrate membrane distillation with low-grade heat producing processes.

Lokare et al. argued in a recent paper that direct contact membrane distillation (DCMD) can have an immense potential in the desalination of highly saline wastewaters where reverse osmosis is not feasible [32]. Their work consisted of evaluating the potential of DCMD for treatment of produced water generated during extraction of natural gas from unconventional reservoirs, otherwise known as shale. The exhaust stream from the natural gas compressor station has been identified as a potential source of waste heat and can be used to operate a DCMD unit. The DCMD unit was shown to be an economically viable option to treat high salinity produced water. They subsequently conducted some ASPEN Plus simulations of DCMD for the desalination of produced/saline water and used laboratory-scale experiments to calibrate the model. The calibrated model was used to optimize the design and operation of theoretical large-scale systems in order to estimate energy requirements for the DCMD process. They determined the optimum membrane area for large scale DCMD plants using the classic concept of minimum temperature approach in heat exchanger design. An energy analysis showed that the waste heat available from natural gas compression station was sufficient to concentrate all the produced water generated in Pennsylvania (USA) to 30 wt% regardless of its initial salinity. This work demonstrates the usefulness of MD in treating high salinity streams toward approaching the so-called zero liquid discharge solution.

Dow et al. [33] operated a direct contact membrane distillation (DCMD) system powered with waste heat from Ecogen Energy's Newport Power Station near Melbourne, Australia. The DCMD system was tested for water recovery from saline demineralization regeneration waste. The source of low-grade waste heat was a gas fired power station that releases heat at less than 40°C and also provided wastewater to the DCMD system fitted with a total membrane area of 0.67 m^2. The trial was operated over a period of 3 months without replacing the membranes or module and achieved 92.8% water recovery. The permeate flux obtained was approximately 3 $L/(m^2 \cdot h)$ and depended solely on the waste heat temperature being supplied. They observed that membrane fouling affected flux and thermal energy demand only toward the end of the trial. The MD system produced a high-quality distillate with an average dissolved solids rejection of 99.9%. Small amounts of ammonia and carbon dioxide were found in the permeate, indicating that MD allows all forms of vapor of components present in the feed to diffuse alongside water vapor. The used membrane autopsy showed that fouling was mainly inorganic scale but some organic matter on the membrane was also detected. Based on the available energy for a continuously operating 500 MW (electricity generation) rated power station, the treatment potential was estimated at up to 8000 m^3/day. The water produced can be theoretically supplied to industrial, residential or agricultural sites.

Jansen et al. [34] used in-house developed MD modules in pilot (3 units tested, one operated in Singapore and two in the Netherlands) and bench scale (30 units tested) tests to demonstrate the potential of producing high quality product water in a single step with little need for water pretreatment and a thermal energy requirement of approximately 520 MJ/m^3 (144 kWh). They also estimated that for large scale applications in the future, using low cost waste heat sources and assuming an energy requirement of 140–230 MJ/m^3 (39–64 kWh), the operational costs could be competitive compared to conventional desalination techniques, such as reversed osmosis and multi-effect distillation. Jansen et al. [34] also described practical problems encountered while operating the MD pilot plants, such as leaks, data losses and minor membrane fouling.

9.3 Solar Energy Sources for Membrane Distillation

Solar heating of water or aqueous solutions has been widely studied and the review of Tiwari et al. on solar desalination on solar energy and solar collectors constitute a good starting point covering a great deal of theory and associated practical issues [12]. Tiwari and Sahota [35] provided a very useful account of the thermodynamics and heat transfer of solar stills and collectors, and gave an insight into the economics of solar energy projects for freshwater production. Tiwari et al. [36] edited a handbook on solar energy and its applications. This book constitutes an invaluable source of theory and data for solar energy applications for desalination, other water heating applications and solar cells (PV).

Chandrashekara and Avadhesh [16] published a review paper where various solar thermal desalination methods such as direct and indirect methods were discussed. The indirect methods were claimed to be preferable for medium and large-scale desalination systems, whereas the direct methods employing the solar stills were said to be more suitable for small-scale systems. Chandrashekara and Avadhesh argued that the performance of the low cost solar stills can be improved with simple modification utilizing various locally available material. They also indicated that low cost stills can be easily and economically fabricated to meet daily needs of fresh drinking water for small households and communities living in islands and remote coastal areas. The solar technology described can also be used for distillation of brackish water populations residing near river banks. Chandrashekara and Avadhesh showed that the system could be suitable for the fluoride removal from waters high in fluoride content well as treatment of arsenic, mercury, cadmium, coliform, virus and bacteria containing waters [16].

Field testing of solar energy for membrane distillation desalination was particularly active in the Mediterranean region and parts of the Middle East. For instance, the European Commission funded a number of desalination projects using membrane distillation in some countries in the Middle East (Jordan [37]) or around the Mediterranean Sea (Italy [38–40], Spain [41,42], Tunisia [43], Egypt [44]).

The Italian regional government funded the Lympha desalination project ("Development of solar-powered systems for water production in remote sites") in Sicily [45]. These countries are known to have seasonal or permanent water stress situations but are blessed with plenty of sunshine in most of the year. Availability of freshwater in coastal resorts of these countries is particularly important for their economy since they are major tourist destinations almost all year round.

Saffarini et al. published an interesting article on solar-powered membrane distillation [46] where they argued that despite the dozen or so SP-MD pilot systems constructed and tested over at least 2 decades, such systems have not been commercialized or implemented on a large scale yet. Their study focused on SP-MD systems that were tested under actual field conditions for several days and were either fully or partially powered by solar energy. The MD systems reported were evaluated in terms of key performance indicators, such as membrane flux, energy consumption, gained output ratio (GOR), performance ratio and the energy recovery scheme applied. The authors also developed a numerical model to show how an important energy efficiency indicator, namely GOR, can be enhanced by certain operating parameters. The results of their study suggested areas of improvement for future SP-MD projects that included using multiple stages to maximize heat recovery within the MD system. However, one has to be realistic, with the low temperatures used in MD in general, prospects and expectations of great efficiencies have to be moderate.

Raluy et al. [41] published a paper in which they described a 5-year experience and data analysis of a solar compact MD demonstration plant installed in the facilities of the Instituto Tecnológico de Canarias (ITC) in Playa de Pozo Izquierdo (Gran Canary Island, Kingdom of Spain). This location is a prime tourist island where water is available mainly through desalination (the average rainfall in the island is around 148 mm/year during a short winter period). The MD system was designed and installed at the end of 2004 and enjoyed funding from the European Commission and other sources; it was operated for a 5 year period. The system had 3 plate collectors with a total area of 6.96 m^2, production capacity of up to 20 L/h, a solar panel to power the pump and other control systems. The MD module was a permeate gap membrane distillation type (PGMD) with a total membrane area of 8.5–10 m^2. The solar collector plates made it possible to reach feed temperatures of 60°C–80°C when performance was optimal and summer operations lasted for 8–9 h. The distillate quality ranged 5 to 695 μS (average 32 μS). They reported a solar irradiance range of 7874–1179 Wh/m^2 (average 6113 Wh/m^2). Obviously, the solar-powered MD systems performance varied according to the time of the day and the season; however, for the best module employed, typical GOR values reported were 05–5.1 (average 3.5) and recovery rate 0.1%–54% (average 18%) and specific heat consumption ranging 138.5–499.1 kWh/m^3 (average 196.4 kWh/m^3).

Schwantes et al. [42] reported in 2013 some work on solar-powered membrane distillation that took place in Gran Canaria, Spain, involving 12 MD modules with a total membrane area of 120 m^2 and a design production capacity of 3.5 m^3/day with seawater feed (35,000 ppm) temperature range 60°C–80°C. The feed flowrate at 80°C was 4800 L/h. The solar collectors used were plates with a total area of 185.6 m^2. They achieved an average GOR of 3.1 when the evaporator/condenser temperatures were 80°C/23°C. This projected was also funded by the European Commission.

More recent work on solar-powered membrane distillation (SPMD) describes advances made to make SPMD more efficient.

Zhang et al. [47] reported a study in which photothermal nanofluids were employed as the feed solution for energy harvesting in solar-powered membrane distillation. Ten different nanofluids were compared and TiN (titanium nitride) was selected following UV-Vis-NIR-waveband (ultraviolet–visible–near-infrared) optical absorption analysis, zeta potential measurement and membrane distillation flux performance. The desalination experiments were conducted using a range of TiN concentrations and solar radiation powers. The results showed that permeate flux and solar energy utilization efficiency increased with increasing TiN content. Compared to the base fluid (35 g/L NaCl aqueous solution), flux increased from 0.47 to 0.74 kg/(m^2·h), while energy utilization efficiency improved from 32.1% to 50.5% for 100 mg/L TiN nanofluid.

The above being an extract of the literature, additional references on SPMD can be found in [48–56].

9.4 Main Challenges in Tapping Low-Grade Heat in Membrane Distillation

Tapping low-grade waste heat from real industrial processes is not a trivial task nor is it within easy reach to any researcher. Permissions and partnerships with the real process plant operator would be required so long as the initiative and work during the approved period will not disrupt the real plant operations nor cause damage to equipment and pipework. In addition, the source of low-grade heat often cannot be utilized directly and would require the use of heat recovery equipment such specialist heat exchangers and means to discard or redirect safely the spent stream after extracting the low-grade heat. Because of these constraints and from the evidence seen in the literature on using industrial low-grade waste heat, the number and nature of processes that can be thermally integrated to MD units will be reduced despite the vast amount of low-grade heat theoretically recoverable. However, if sufficient experience is gained on using low-grade waste heat from industrial sources, the actual owners or operators of such plants can "plan" for thermally integrated MD units at the design stage of future processes instead of using existing means to discharge waste heat into the environment.

9.5 Main Challenges in Tapping Solar Energy in Membrane Distillation

Tapping solar energy to power MD units normally should be less constrained than tapping into industrial low-grade waste heat. However, given the transient and discontinuous nature of solar energy, the infrastructure of the solar energy harvesting can be as costly as the membrane distillation unit or more. Unfortunately we do not currently have cost breakdown for solar-powered membrane distillation (SPMD) units. However, one can easily envision a number of scenarios where low cost, unsophisticated SPMD units can be designed for low budget cases where reliability and optimum performance is not critical. Or, on the other hand, very sophisticated SPMD units can be designed for optimum performance and high reliability of operations. The latter scenario would probably be the option for producing ultra pure water for high value applications.

The sophistication could be found in solar irradiance tracking, flow and temperature control, and heat loss minimization.

9.6 Concluding Remarks

Membrane distillation for desalination using low-grade waste heat and solar energy has a great potential that is yet to be demonstrated on a large scale. Even though we saw that the number of papers published on MD for desalination is growing fast, we have yet to see hard data on energy consumption and cost of fresh water produced using a commercial scale MD plant. The review paper of Khayet [25] indicated that there is a scatter of data on energy consumption and production cost of freshwater using MD systems. This is perhaps indicative of the real situation where in the absence of fully optimized, tested and deployed full scale MD stations, it becomes unrealistic to talk about reliable power consumption and water production cost. Suffice it to say that when membrane distillation is thermally integrated with process plants that currently discharge low-grade heat to the environment, there is scope for utilizing the low-grade heat (that would otherwise be lost) to produce high quality process water, thus making savings that can at least partially offset costs of waste heat recovery equipment. In the case of using waste heat from ship engines, few studies were made and they were often based on simulation of heat source, except for the paper of Xu et al. [27] where MD tests were conducted onboard a real ship. Such application would qualify as "niche" application and can be very beneficial given the high quality of freshwater produced. Such water can be used for drinking (after mineralization), for hygiene or in the engine room as process water.

It is, however, imperative to conduct studies on the potential risks to the "parent" process if the MD unit fails or shuts down, so that the waste heat can be safely discharged without upsetting the main process.

On the other hand, solar-powered membrane distillation (SPMD) has been tested on pilot plant scale in several European and Middle Eastern countries. The most recent water production costs of a pilot plant scale SPMD ranged 10–11.30 Euros/m^3 [56]. While this is considerably higher than the cost of production of water from thermal or reverse osmosis plants (quoted to be in the range 0.5 to 1 \$/m^3 [57]), it has to be remembered that MD will probably not compete with thermal or reverse osmosis desalination on a cost of production basis, but will compete on a convenience basis for remote sites and for high value industries requiring ultra-pure water (process water, pharmaceuticals for instance). In the end, whatever the MD feed heat source, be it solar or low-grade waste heat, the technological breakthrough in MD can only be accomplished if the membrane resistance to fouling and wetting is demonstrated or fouling and wetting will be greatly minimized by using low cost anti scaling additives.

References

1. Elimelech, M. and W.A. Phillip, The future of seawater desalination: Energy, technology, and the environment. *Science*, 2011. **333**(6043): pp. 712–717.
2. Raluy, G., L. Serra, and J. Uche, Life cycle assessment of MSF, MED and RO desalination technologies. *Energy*, 2006. **31**(13): pp. 2361–2372.
3. Schneider, K. et al., Membranes and modules for transmembrane distillation. *Journal of Membrane Science*, 1988. **39**(1): pp. 25–42.
4. Kelly, R.C., *The Carbon Conundrum: Global Warming and Energy Policy in the Third Millennium*. 2002, Houston, TX: CountryWatch.
5. Forman, C. et al., Estimating the global waste heat potential. *Renewable and Sustainable Energy Reviews*, 2016. **57**: pp. 1568–1579.
6. Law, R., A. Harvey, and D. Reay, A knowledge-based system for low-grade waste heat recovery in the process industries. *Applied Thermal Engineering*, 2016. **94**: pp. 590–599.
7. Rattner, A., US Waste Heat Resources. 2018 [December 9, 2018]; Available from: http://sites.psu.edu/mtfe/us-waste-heat-resources/
8. Rahimi, B. et al., A novel process for low grade heat driven desalination. *Desalination*, 2014. **351**: pp. 202–212.
9. Ammar, Y. et al., Desalination using low grade heat in the process industry: Challenges and perspectives. *Applied Thermal Engineering*, 2012. **48**: pp. 446–457.
10. Shih, H. and T. Shih, Utilization of waste heat in the desalination process. *Desalination*, 2007. **204**(1–3): pp. 464–470.
11. Talbert, S.G., J.A. Eibling, and G.O.G. Lof, Manual on solar distillation of saline water, U.D.o.t. I. R&D Progress Report No. 546, Editor. 1970: US Department of the Interior.
12. Tiwari, G.N., H.N. Singh, and R. Tripathi, Present status of solar distillation. *Solar Energy*, 2003. **75**(5): pp. 367–373.
13. Pugsley, A. et al., Chapter 2 – Solar desalination potential around the world, eds. V.G. Gude, *Renewable Energy Powered Desalination Handbook*. 2018, Amsterdam, the Netherlands: Butterworth-Heinemann, pp. 47–90.
14. Cen, J. et al., Experimental study on a direct water heating PV-T technology. *Solar Energy*, 2018. **176**: pp. 604–614.
15. Sharshir, S.W. et al., Thermal performance and exergy analysis of solar stills – A review. *Renewable and Sustainable Energy Reviews*, 2017. **73**: pp. 521–544.
16. Chandrashekara, M. and A. Yadav, Water desalination system using solar heat: A review. *Renewable and Sustainable Energy Reviews*, 2017. **67**: pp. 1308–1330.
17. Kaviti, A.K., A. Yadav, and A. Shukla, Inclined solar still designs: A review. *Renewable and Sustainable Energy Reviews*, 2016. **54**: pp. 429–451.
18. Prakash, P. and V. Velmurugan, Parameters influencing the productivity of solar stills – A review. *Renewable and Sustainable Energy Reviews*, 2015. **49**: pp. 585–609.
19. Sharaf, M.A., Thermo-economic comparisons of different types of solar desalination processes. *Journal of Solar Energy Engineering-Transactions of the ASME*, 2012. **134**(3): p. 031001.
20. Compain, P., Solar energy for water desalination. *Procedia Engineering*, 2012. **46**: pp. 220–227.

21. Kabeel, A.E. and S.A. El-Agouz, Review of researches and developments on solar stills. *Desalination*, 2011. **276**(1): pp. 1–12.
22. Stefan, P. and S. Iain, The Renewables.ninja . 2018 [cited December 9, 2018]; Available from: https://www.renewables.ninja/
23. Sharshir, S.W. et al., Factors affecting solar stills productivity and improvement techniques: A detailed review. *Applied Thermal Engineering*, 2016. **100**: pp. 267–284.
24. Qtaishat, M.R. and F. Banat, Desalination by solar powered membrane distillation systems. *Desalination*, 2013. **308**: pp. 186–197.
25. Khayet, M., Solar desalination by membrane distillation: Dispersion in energy consumption analysis and water production costs (a review). *Desalination*, 2013. **308**: pp. 89–101.
26. Gude, V.G., Use of exergy tools in renewable energy driven desalination systems. *Thermal Science and Engineering Progress*, 2018. **8**: pp. 154–170.
27. Xu, Y., B.K. Zhu, and Y.Y. Xu, Pilot test of vacuum membrane distillation for seawater desalination on a ship. *Desalination*, 2006. **189**(1–3): pp. 165–169.
28. Amaya-Vías, D., E. Nebot, and J.A. López-Ramírez, Comparative studies of different membrane distillation configurations and membranes for potential use on board cruise vessels. *Desalination*, 2018. **429**: pp. 44–51.
29. González, D., J. Amigo, and F. Suárez, Membrane distillation: Perspectives for sustainable and improved desalination. *Renewable and Sustainable Energy Reviews*, 2017. **80**: pp. 238–259.
30. Long, R. et al., Direct contact membrane distillation system for waste heat recovery: Modelling and multi-objective optimization. *Energy*, 2018. **148**: pp. 1060–1068.
31. Lai, X. et al., A hybrid system using direct contact membrane distillation for water production to harvest waste heat from the proton exchange membrane fuel cell. *Energy*, 2018. **147**: pp. 578–586.
32. Lokare, O.R. et al., Integrating membrane distillation with waste heat from natural gas compressor stations for produced water treatment in Pennsylvania. *Desalination*, 2017. **413**: pp. 144–153.
33. Dow, N. et al., Pilot trial of membrane distillation driven by low grade waste heat: Membrane fouling and energy assessment. *Desalination*, 2016. **391**: pp. 30–42.
34. Jansen, A.E. et al., Development and pilot testing of full-scale membrane distillation modules for deployment of waste heat. *Desalination*, 2013. **323**: pp. 55–65.
35. Tiwari, G.N. and L. Sahota, Chapter 14 – Exergy and technoeconomic analysis of solar thermal desalination, eds. V.G. Gude, *Renewable Energy Powered Desalination Handbook*. 2018, Amsterdam, the Netherlands: Butterworth-Heinemann, pp. 517–580.
36. Tiwari, G.N., Shyam, and A. Tiwari, Handbook of Solar Energy Theory, Analysis and Applications. 2016, Singapore: Springer.
37. Banat, F. et al., Desalination by a "compact SMADES" autonomous solarpowered membrane distillation unit. *Desalination*, 2007. **217**(1–3): pp. 29–37.
38. Cipollina, A., G. Micale, and L. Rizzuti, Membrane distillation heat transfer enhancement by CFD analysis of internal module geometry. *Desalination and Water Treatment*, 2011. **25**(1–3): pp. 195–209.
39. Criscuoli, A., M.C. Carnevale, and E. Drioli, Modeling the performance of flat and capillary membrane modules in vacuum membrane distillation. *Journal of Membrane Science*, 2013. **447**: pp. 369–375.

40. Tamburini, A. et al., A thermochromic liquid crystals image analysis technique to investigate temperature polarization in spacer-filled channels for membrane distillation. *Journal of Membrane Science*, 2013. **447**: pp. 260–273.
41. Raluy, R.G. et al., Operational experience of a solar membrane distillation demonstration plant in Pozo Izquierdo-Gran Canaria Island (Spain). *Desalination*, 2012. **290**: pp. 1–13.
42. Schwantes, R. et al., Membrane distillation: Solar and waste heat driven demonstration plants for desalination. *Desalination*, 2013. **323**: pp. 93–106.
43. MEDIRAS Project, E.C. *Seawater compact system in Tunisia*. 2011 [cited October 10, 2018]; http://www.mediras.eu/index.php@id=116.html
44. Fath, H.E.S. et al., PV and thermally driven small-scale, stand-alone solar desalination systems with very low maintenance needs. *Desalination*, 2008. **225**(1): pp. 58–69.
45. Cipollina, A. et al., Development of a membrane distillation module for solar energy seawater desalination. *Chemical Engineering Research & Design*, 2012. **90**(12): pp. 2101–2121.
46. Saffarini, R.B. et al., Technical evaluation of stand-alone solar powered membrane distillation systems. *Desalination*, 2012. **286**: pp. 332–341.
47. Zhang, Y. et al., Enhancement of energy utilization using nanofluid in solar powered membrane distillation. *Chemosphere*, 2018. **212**: pp. 554–562.
48. Soomro, M.I. and W.-S. Kim, Performance and economic evaluation of linear Fresnel reflector plant integrated direct contact membrane distillation system. *Renewable Energy*, 2018. **129**: pp. 561–569.
49. Soomro, M.I. and W.-S. Kim, Parabolic-trough plant integrated with direct-contact membrane distillation system: Concept, simulation, performance, and economic evaluation. *Solar Energy*, 2018. **173**: pp. 348–361.
50. Dongare, P.D. et al., Nanophotonics-enabled solar membrane distillation for off-grid water purification. *Proceedings of the National Academy of Sciences*, 2017. **114**(27): pp. 6936–6941.
51. Suárez, F. and R. Urtubia, Tackling the water-energy nexus: An assessment of membrane distillation driven by salt-gradient solar ponds. *Clean Technologies and Environmental Policy*, 2016. **18**(6): pp. 1697–1712.
52. Shim, W.G. et al., Solar energy assisted direct contact membrane distillation (DCMD) process for seawater desalination. *Separation and Purification Technology*, 2015. **143**: pp. 94–104.
53. Summers, E.K. and J.H. Lienhard, Experimental study of thermal performance in air gap membrane distillation systems, including the direct solar heating of membranes. *Desalination*, 2013. **330**: pp. 100–111.
54. Abdallah, S.B., N. Frikha, and S. Gabsi, Simulation of solar vacuum membrane distillation unit. *Desalination*, 2013. **324**: pp. 87–92.
55. Kabeel, A.E., M. Abdelgaied, and E.M.S. El-Said, Study of a solar-driven membrane distillation system: Evaporative cooling effect on performance enhancement. *Renewable Energy*, 2017. **106**: pp. 192–200.
56. Guillén-Burrieza, E. et al., Techno-economic assessment of a pilot-scale plant for solar desalination based on existing plate and frame MD technology. *Desalination*, 2015. **374**: pp. 70–80.
57. Zhou, Y. and R.S.J. Tol, Evaluating the costs of desalination and water transport. *Water Resources Research*, 2005. **41**(3).

10

Conclusions and Future Horizons for Membrane Distillation Desalination

10.1 Introduction

One of the trends noticed recently in review papers and possibly in other forms of publication was to compile the number of times "membrane distillation" is cited in scientific databases and the count of how many papers were published on "membrane distillation" and what type of configuration, etc. While this may show the growing interest in this technology that is not strictly new, because it dates back from 1960s and was put on hold for a while until the early 1990s perhaps when interest was renewed; what was not quoted was the amount of overlap in information in multiple journals and the amount of data scatter (flux values, energy consumption, etc.). It is perhaps time to consolidate efforts in MD development toward making a leap in technological advances that will make commercial deployment low cost in remote underprivileged communities that have been deprived of their human right access to freshwater (Resolution 64/292 of the United Nations General Assembly) [1]. It is quite staggering to notice that only a handful of pilot plant trials have been conducted in the past two decades [2].

Let us review some of the issues that may need to be overcome in the future to help make that leap toward deployment of MD where it matters most.

10.2 Outstanding Issues That Hinder Commercial Deployment of Membrane Distillation for Desalination

There is a need to invest efforts in solar-powered membrane distillation (SPMD) in underprivileged communities that are considered water stressed to relieve the hardship of people walking for miles to fetch water, sometimes polluted for day to day needs. Here, we have an issue of will from the so-called G20 states who can afford to assist financially as part of their overseas development and aid programs. There was a first step in 2014 for the deployment of a vacuum MD station (MEMSYS brand) [3] in the Maldives exploiting low-grade waste heat from the exhaust of a diesel generator. "The pricing of the water is very moderate to make it affordable to the population but covering the costs of the plant and operations" [4]. Unfortunately, no figure for the cost was quoted, and yet the cost of water in underprivileged parts of the world remains a barrier.

Membrane fouling continues to be a subject of research, and there has been a massive growth in improved membrane research programs [5–14]. Without a major leap in membrane development to overcome the problem of membrane fouling, there will be a risk of slow or lack of deployment of MD for desalination in parts of the world where membrane change and maintenance may not be economically sustainable. The cost of future improved membranes should not be a major fraction of the MD system. Since there are no reliable cost data on MD systems, the nearest reference shows that membrane replacement can be costly [15]. Obviously more work on MD economics is needed using price quotation of material and services rather than old fashioned empirical estimations. A start for improvement in MD economic evaluation was recently reported [16] and needs to be validated further.

10.3 Cost Competitivity Issues

We still do not have accurate costs for pilot plant MD stations (capital cost breakdown and operational costs). Much of the published data on MD production costs are based on assumptions in earlier studies [17] and referenced in later studies as if they were justified [18]. Until there are real commercial plants and costs are disclosed, the current approach to costing MD production remains at best "indicative" and probably inaccurate. In addition, the lessons we learned from reverse osmosis production costs should serve as a guide to how the costs evolve: in the early days of RO, unit production costs were quite high, but they decreased considerably thanks to the development of two critical items in the RO plant: the gold standard "thin film composite" membrane and pressure exchangers. The cost of water produced by large scale RO is nowadays less than $1/m^3 [15]. We have not reached that level in membrane development nor in energy consumption competitivity in membrane distillation desalination. In fact the estimated cost of water produced by renewable energy powered MD were quoted to be in the range $13–$18/m^3 [15]. This price range is not far from "low-tech" solar stills that were quoted to be in the range $12–$12.5 [15]. In a recent review paper, the cost showed a tremendous scatter ranging from $0.6 to $18/m^3 [2]. The question we can ask legitimately would be what makes MD production costs so high with SPMD? Unfortunately, not many deep studies were conducted and one of the most cited sources indicated that it is not the cost of membrane, but the "rest of the MD" plant, namely the membrane module and support, instrumentation and monitoring, heat exchangers, solar collectors (or waste heat recovery devices). There is little research on bringing the cost of these down, perhaps with the exception of solar collectors where prices came down recently (mainly due to an interest in solar water heating in smart homes).

10.4 Sustainability Issues of Membranes for Membrane Distillation

Water is an essential commodity that humans cannot do without. The global population growth and its implications for demand for water has been highlighted in Chapter 1. This growing demand for water is now compounded by the water scarcity issue in a growing number of regions with an acute water stress situation, as shown in Chapter 1. Seawater desalination remains the only sustainable solution from the point of view of feed to desalination facilities. Oceans and seas will not deplete any time soon. The real concern is the energy requirements to desalinate seawater or brackish water to produce freshwater. Currently, reverse osmosis is the desalination technology of choice for energy efficiency: the best RO available these days require a total of 3.57 kWh/m^3 in which 71% of the energy consumed goes toward overcoming the osmotic pressure and membrane resistance; the rest would be consumed elsewhere in the plant (Tables 1.4 and 1.5). On the other hand, energy input in MD pilot plants requires 39–64 kWh/m^3 [19]. This is at least 10 times the energy consumed in RO. While this may not be a major issue when solar power or waste heat are used (both are theoretically free), the issue would be in the infrastructure cost to harvest that much energy. When the SPMD or waste heat MD plants are scaled up, what would be the largest scale affordable and what would be the corresponding production capacity of freshwater? What would be the priority for utilization of that freshwater produced? Such questions need to be answered in the context of socio-economics and will require some profound thought.

10.5 Target Applications of Membrane Distillation for Desalination

We can envisage two types of applications where deployment can be made soon and where production capacity is not so high that the capital cost of the MD module and associated ancillary equipment would be prohibitive. First, one can think of thermally integrating MD plants with process plants where low-grade waste heat is rejected. The obvious application would be production of high purity process water (assuming the plant is located near the sea). Substantial savings can be made to bypass ion exchange resin units to produce deionized water and saving the cost of purchase of freshwater or eliminating small desalination plants in the parent plant if the latter is located in arid regions. This application would be ideal because the technical know-how of personnel in the host process plant will ensure the MD unit will be well looked after in terms of maintenance and repairs if needed. We anticipate a long productive life of the MD unit if it were thermally integrated in a process plant.

A second type of MD application would be a SPMD installed in a small coastal community isolated from the water distribution grid. There are also many islands where freshwater is only available through tanker ships. For these communities, SPMD in association with photovoltaic electricity generation to drive pumps and instrumentation would be perfectly feasible. There have been pilot trials in Spain [20] and Singapore [21] on SPMD and some experience was gained. Future development for more permanent deployment can address issues related to the transient nature of solar heating and its effect on the membrane life. SPMD has probably the best prospects for deployment in remote coastal regions or regions with saline aquifers. The quality of life in such regions would improve dramatically for the better indeed.

10.6 Emergence of Computational Fluid Dynamics in Membrane Distillation Modeling

Module design is a key stage in MD development and can be costly if fabrication errors were made. It was recommended in Chapter 4 on membrane distillation module design to resort to computational fluid dynamics (CFD) to explore a number of mechanical designs before committing to final workshop fabrication. There is a need for more elaborate CFD models to relax simplifying assumptions that were often introduced to make simulations run faster or if the computing resources are modest. In addition, there is also a need to make MD phenomena built-in CFD packages as standard to enable a wider participation of researchers.

References

1. United Nations General Assembly. The human right to water and sanitation. 2010 [cited December 15, 2018]; Available from: http://www.un.org/en/ga/search/view_doc.asp?symbol=A/RES/64/292.

2. Thomas, N. et al., Membrane distillation research & implementation: Lessons from the past five decades. *Separation and Purification Technology*, 2017. **189**: pp. 108–127.
3. Memsys, drinking water paradise found: Maldives drinking water is produced from waste heat. *Filtration & Separation*, 2014. **51**(3): pp. 44–45.
4. Water&WastesDigest. Aquiva Foundation marks two-year anniversary of desalination plant in Maldives. 2016 [cited December 15, 2018]; Available from: https://www.wwdmag.com/membrane-technology/aquiva-foundation-marks-two-year-anniversary-desalination-plant-maldives.
5. Zhu, Z. et al., Dual-bioinspired design for constructing membranes with superhydrophobicity for direct contact membrane distillation. *Environmental Science & Technology*, 2018. **52**(5): pp. 3027–3036.
6. Zheng, R. et al., Preparation of omniphobic PVDF membrane with hierarchical structure for treating saline oily wastewater using direct contact membrane distillation. *Journal of Membrane Science*, 2018. **555**: pp. 197–205.
7. Zhao, J. et al., Preparation of PVDF/PTFE hollow fiber membranes for direct contact membrane distillation via thermally induced phase separation method. *Desalination*, 2018. **430**: pp. 86–97.
8. Yang, H.-C. et al., Janus membranes: Creating asymmetry for energy efficiency. *Advanced Materials*, 2018. **30**(43): p. 1801495.
9. Yang, F. et al., Metal–organic frameworks supported on nanofiber for desalination by direct contact membrane distillation. *ACS Applied Materials & Interfaces*, 2018. **10**(13): pp. 11251–11260.
10. Wang, M. et al., ZnO nanorod array modified PVDF membrane with superhydrophobic surface for vacuum membrane distillation application. *ACS Applied Materials & Interfaces*, 2018. **10**(16): pp. 13452–13461.
11. Sun, W. et al., An ultrathin, porous and in-air hydrophilic/underwater oleophobic coating simultaneously increasing the flux and antifouling property of membrane for membrane distillation. *Desalination*, 2018. **445**: pp. 40–50.
12. Qiu, H. et al., Pore channel surface modification for enhancing anti-fouling membrane distillation. *Applied Surface Science*, 2018. **443**: pp. 217–226.
13. Mejia Mendez, D.L. et al., Membrane distillation (MD) processes for water desalination applications: Can dense selfstanding membranes compete with microporous hydrophobic materials? *Chemical Engineering Science*, 2018. **188**: pp. 84–96.
14. Lu, K.J. et al., Omniphobic hollow-fiber membranes for vacuum membrane distillation. *Environmental Science & Technology*, 2018. **52**(7): pp. 4472–4480.
15. Banat, F. and N. Jwaied, Economic evaluation of desalination by small-scale autonomous solar-powered membrane distillation units. *Desalination*, 2008. **220**(1–3): pp. 566–573.
16. Hitsov, I. et al., Economic modelling and model-based process optimization of membrane distillation. *Desalination*, 2018. **436**: pp. 125–143.
17. Al-Obaidani, S. et al., Potential of membrane distillation in seawater desalination: Thermal efficiency, sensitivity study and cost estimation. *Journal of Membrane Science*, 2008. **323**(1): pp. 85–98.
18. Kesieme, U.K. et al., Economic analysis of desalination technologies in the context of carbon pricing, and opportunities for membrane distillation. *Desalination*, 2013. **323**: pp. 66–74.

19. Jansen, A.E. et al., Development and pilot testing of full-scale membrane distillation modules for deployment of waste heat. *Desalination*, 2013. **323**: pp. 55–65.
20. Guillén-Burrieza, E. et al., Experimental analysis of an air gap membrane distillation solar desalination pilot system. *Journal of Membrane Science*, 2011. **379**(1): pp. 386–396.
21. Zhao, K. et al., Experimental study of the memsys vacuum-multi-effect-membrane-distillation (V-MEMD) module. *Desalination*, 2013. **323**: pp. 150–160.

Index

Note: Page numbers in italic and bold refer to figures and tables, respectively.